T0257769

Ultra-Wideband Radio Technologies

Ultra-Wideband Radio Technologies

Edited by **Kevin Merriman**

CLANRYE INTERNATIONAL

New Jersey

Published by Clanrye International,
55 Van Reypen Street,
Jersey City, NJ 07306, USA
www.clanryeinternational.com

Ultra-Wideband Radio Technologies
Edited by Kevin Merriman

© 2015 Clanrye International

International Standard Book Number: 978-1-63240-504-3 (Hardback)

This book contains information obtained from authentic and highly regarded sources. Copyright for all individual chapters remain with the respective authors as indicated. A wide variety of references are listed. Permission and sources are indicated; for detailed attributions, please refer to the permissions page. Reasonable efforts have been made to publish reliable data and information, but the authors, editors and publisher cannot assume any responsibility for the validity of all materials or the consequences of their use.

The publisher's policy is to use permanent paper from mills that operate a sustainable forestry policy. Furthermore, the publisher ensures that the text paper and cover boards used have met acceptable environmental accreditation standards.

Trademark Notice: Registered trademark of products or corporate names are used only for explanation and identification without intent to infringe.

Printed in the United States of America.

Contents

Preface

This book provides an in-depth analysis of the technology and applications of Ultra-Wideband Radio (UWB). It includes an analysis of the applications of UWB technologies in various fields. Ultra-Wideband Radio, referred to as UWB, is a characteristic of a new radio access philosophy and utilizes several GHz of bandwidth. It promises increased data rate communication over small distances, as well as modern radar sensing and localization applications with extraordinary resolution. Fields of function may be found in industry, civil engineering, observation and examination for security and safety measures, and even for medicine. The book compiles researches conducted by prominent experts and will be useful for readers interested in this subject.

The information contained in this book is the result of intensive hard work done by researchers in this field. All due efforts have been made to make this book serve as a complete guiding source for students and researchers. The topics in this book have been comprehensively explained to help readers understand the growing trends in the field.

I would like to thank the entire group of writers who made sincere efforts in this book and my family who supported me in my efforts of working on this book. I take this opportunity to thank all those who have been a guiding force throughout my life.

Editor

ISOPerm: Non-Contacting Measurement of Dielectric Properties of Irregular Shaped Objects

Henning Mextorf, Frank Daschner, Mike Kent and Reinhard Knöchel

Additional information is available at the end of the chapter

1. Introduction

A mere glance at the contents of any of the conferences organised by ISEMA (International Society for Electromagnetic Aquametry) [1] shows that the measurement and control of water (its quantity and states) in materials is a very wide and active field. That water lends itself to study in this way is because of its very dominant dispersive dielectric properties and an unusually large dipole moment for such a small molecule (1.84 Debye units). At room temperature the dispersion is centred on $\sim$ 12.5GHz and the real part of the relative permittivity at its upper and lower frequency extremes is $\sim$ 4.3 and 80 respectively [2]. The complex dielectric properties characteristic of this dispersion are the properties that are measured and correlated with whatever aspect of the water content is of interest. Historically much of the work was carried out at one or two frequencies, mostly in X-band where the dielectric loss is at a maximum, but the advent of time domain reflectometry (TDR) [3] for broadband dielectric measurements in the microwave region eventually led to such measurements being made using open ended coaxial sensors [4], although a greater potential of such measurements was only realised later by the authors. The use of such sensors freed the experimenter from the difficult task of defining the sample shape by means of a sample cell; the measurements still required however that the sensor be in contact with the sample. As a true frequency domain UWB application, dielectric measurements of foodstuffs over a wide range of frequencies (100MHz to 20GHz) were made using network analysers and such coaxial sensors [5, 6]. Drawing on the experience in other chemometric applications such as NIR (near infra-red spectroscopy) [7], the dielectric spectra obtained were subjected to various multivariate analyses (PCR (principal component regression), PLSR (partial least squares regression), and ANNs (artificial neural networks both linear and non-linear)). Such analyses both compress the data into orthogonal factors and extract from those factors the best information to predict the composition of the foodstuffs. In such analysis the important variables are, not so much the dielectric properties at each sampled frequency, but rather the shape of the spectrum. In effect the data reduction provides suitable shape descriptors, which are in the case of foods, very dependent on the water content and its interaction with other constituents such as proteins and carbohydrates. Any that are dependent on other factors are eliminated in the regression analysis having no significant correlation with the material properties. Rather than using measurements in the frequency domain, of course it is possible to transform time domain measurements to the frequency domain using Fourier or other forms of transformation. For time varying data acquired by TDR the inverse Fourier transform in its most general form can be written as in equation 1.

$$h(t) = \frac{1}{2\pi} \int_{-\infty}^{+\infty} g(f)e^{2\pi i f t} \mathrm{d}f, \tag{1}$$

where h(t) is a time dependent function the Fourier transform of which is a frequency dependent spectrum g(f). Examination of equation (1) shows that at any instant t every component part of the spectrum contributes to the value of h. Because the subsequent multivariate analysis can be thought of as dealing with shape and is concerned only with the variations in g then transformation of h(t) to the frequency domain is not required, since related variations are present in h(t) and the shape of the time domain function is equally useful. A further justification for eliminating the transformation step concerns the difficulties with which it is fraught. Firstly, the truncation of the pulse after a finite measurement interval can introduce undesirable distortions in the integral from convolution of the pulse with the rectangular time window (windowing). Secondly, the act of sampling the pulse at regular intervals means that frequencies present with periods shorter than the sampling interval are incorporated as lower frequency information (aliasing), and thirdly, to accurately reconstruct the reflection coefficients in the frequency domain exact time referencing of the pulse is required, else phase errors cause large inaccuracies at the high frequency end of the spectrum. By carrying out the analysis on the raw, sampled TDR pulse, all the errors above can be avoided. This was tried and demonstrably gave the same results as the spectral data, with less computer effort and fewer error generating problems. The first problem to which this was applied was something less tangible than water content: it was in fact the quality of various seafoods as defined by more subjective methods [8]. The success of this approach naturally led to attempts to measure dielectric objects in a non-contacting fashion, using firstly transmitted UWB quasi-Gaussian pulses of 400ps width [9]. The transmitting and receiving antennas in this initial work were double ridged horns and the sample was arranged in a wide layer of uniform thickness. In such a situation, additional interfering variables can be the position of the antennae, polarization of the transmitted wave, multiple path effects and a host of others, all of which may be eliminated by multivariate analysis. Effects due to dielectric properties can be separated from those due to geometry and other exterior factors. Thereafter, further work followed with samples of increasingly complex shapes and different orientations, beginning with simple rectangles and progressing to other shapes such as triangles and circles [10, 11], albeit still of constant thickness. At the same time, various forms of UWB antennas were investigated but currently the choice is an array of simple dipoles for the receiving antennas with a horn antenna transmitting the pulse. The multivariate data analysis still uses PCA as a first step to reduce the data and provide shape descriptors, but because PCR is not entirely suitable for non-linear processes (being a linear regression of variables) the PCs are used as input to a non-linear ANN. A great deal of work has now been done, gradually broadening the application parameters until now it is possible to measure UWB dielectric properties of objects with any shape, thickness, orientation and without contact [12-20]. This has been the subject of the project 'ISOPerm' (irregularly shaped objects-permittivity) the methods and results of which will now be described.

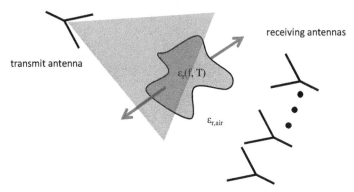

Figure 1. Irregularly shaped object illuminated by an electromagnetic wave.

2. Problem and approach

There is no appropriate method for the determination of the dielectric or related properties of irregular shaped objects in free-space. The investigated objects are considered to be small compared to the range of wavelengths and the footprints of the antennas used. A visualization of the problem is depicted in Figure 1. An electromagnetic wave illuminates an irregularly shaped object with its frequency and temperature dependent dielectric properties. It is surrounded by air. Portions of the scattered field are collected by multiple field probes, i.e. a line array of receiving antennas. The scattered signals contain information about geometry as well as the dielectric properties. Because of the complexity of the problem it is assumed that the development of a physical model (as used in conventional free-space methods having a plane parallel plate) would be too complex. Furthermore, an on-line method suitable for characterizing many objects in a short time is anticipated. Therefore, multivariate calibration methods are applied in order to separate dielectric from geometric influences. The objects are considered to be homogeneous and non-magnetic ($\mu_r = 1$). However, it is assumed that it would also be possible to predict both average permittivity and permeability. In research and industrial applications other properties, for example the water content, moisture, freshness or quality of foodstuffs, are of great interest. They are often strongly correlated to dielectric properties and can be determined directly without knowing the permittivity.

3. Measurement system and dedicated hardware

A measurement system has been built in order to verify the approach. Particular components, like the UWB-antennas used and the working principle of the whole set-up, are discussed in this section. A dedicated time domain transmission oscilloscope is presented. It is capable of transmitting and receiving UWB pulses with several gigahertz of bandwidth. It is specially tailored to the measurement problem and therefore functions with less hardware and software, is far more compact, and is cheaper compared to a universal laboratory instrument.

3.1. Compact ultra-wideband antennas and arrays

The sensor system should be able to transmit and receive ultra-wideband (UWB) signals in two orthogonal polarizations. This is of great importance if the orientation of the object

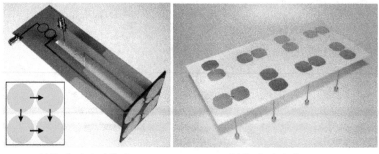

(a) Dual polarised dipole.   (b) Dual polarised array.

Figure 2. Ultra-wideband antenna structures.

under test (e.g. prolate ellipsoid) is not known. There are several polarization set-ups for receiving co- and cross-polarized parts of the scattered wave. In the set-up used latterly the polarization of the transmitting antenna is tilted at 45° with respect to the receiving array so that ideally, without an object under test, the receiving antennas receive the transmitted pulse in equal proportions for each polarization. Therefore, dual orthogonal UWB antennas have been developed. Furthermore the single antennas should be as compact as possible while operating in the lower gigahertz range. The arrays presented later are not group antennas as commonly used. The received signal of every antenna is sampled separately in order not to confound the information at different locations with respect to the object under test. Operating at frequencies in the lower GHz range is anticipated because the electromagnetic waves have a higher penetration depth (especially for lossy dielectrics with a high permittivity) and therefore will be more affected.

There are many UWB antennas suitable for building dual polarized antennas [21–25]. Dual polarized dipole (see [26–31]) and horn antennas (see [32–35]) can already be found in the literature. Most of these antennas have the disadvantage that they have crossed feed points. It is possible to overcome this problem with the configuration depicted in Figure 2. Both the four radiation elements and the whole configuration are two-fold symmetric. Two dipoles are excited in even mode in horizontal as well as in vertical polarization. Ideally, due to the symmetry, both planes are decoupled. The diameter of the radiation elements is 24mm; the whole PCB containing the four radiating elements measures 50mm × 50mm. The antenna is equipped with two orthogonal feed networks; each is composed of a two stage Wilkinson divider and two tapered baluns. The horizontal and vertical polarization feed networks have dimensions of 50mm × 142mm and 50mm × 103mm, respectively. In order to avoid backward radiation, absorber material is mounted appropriately (the absorber material was removed for the photograph shown in Figure 2).

The matching at both ports is better than 7dB in a frequency range from 2GHz to 5.7GHz; the isolation is better than 30dB. At 5GHz the radiated crosspolar portion is suppressed by more than 25dB (boresight); the gain is 5.2dB. Measurements with a line containing four of these antennas lead to excellent results but the whole antenna structure is rather complex. Since a scenario with an object moving orthogonally to the line array is anticipated (e.g. a conveyor belt), a simpler arrangement can be used. The antenna array shown in Figure 2 consists of eight single linear broadband dipoles. The geometry is optimized regarding the antenna matching. The radiating elements have a maximum diameter of 30mm. The feed is provided through a coaxial-slotline transition, so that no other components are necessary. The distance between two dipoles having an equal polarization is 80mm in each direction; between dipoles with the same polarization it is $\sqrt{2} \times 80$mm. The dimensions are overall 160mm × 320mm × 100mm. The matching is better than 10dB between 1.6GHz and 4.2GHz. The crosstalk between the single elements is maximum −20dB at 10GHz.

3.2. Signal generation and sampling

The core of the measurement system is the signal generation and sampling. The requirements for the system are that the step signals generated have a rise time in the range of 100ps, and a large amplitude. These signals are transmitted through the measurement path after which, the step response has to be sampled. The system should have a high bandwidth, low noise and low jitter. Furthermore it should be as compact and affordable as possible. Therefore, an impulse technique using equivalent time sampling is the method of choice. Classical swept sine wave techniques, as used in network analysers or real time digital sampling oscilloscopes, are too expensive, complex, and bulky.[1]

Equivalent time sampling is based on the repetitive stimulation by a measurement signal with a cycle duration of T_0. At every repetition of the measurement signal the moment of sampling is shifted by ΔT. The cycle of the sampling clock is then $T_1 = T_0 + \Delta T$. The sampled signal is therefore stretched to $T_2 = \frac{T_0}{\Delta T} T_1$. The effort regarding analog-to-digital conversion and the data transport and storage is greatly reduced compared to a real time oscilloscope. This technique has been used for decades; with recently available MMICs, both high performance and cheap hardware can be achieved. It is the method of choice for the sampling of signals with huge bandwidth in the GHz range with a high resolution. Laboratory instruments employing this technique have been used for preliminary investigations. In order to demonstrate the system performance and accuracy under practical conditions, dedicated hardware was developed. Two specially tailored systems were investigated. They differed

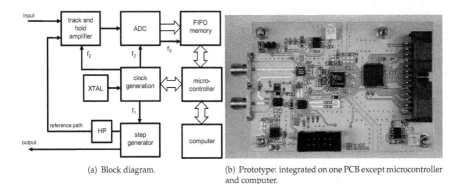

(a) Block diagram. (b) Prototype: integrated on one PCB except microcontroller and computer.

Figure 3. Time domain transmission measurement system with a microcontroller.

---

[1] A rather exotic concept employing M-sequences can be found in this book (see chapter HALOS). There are also measurements carried out with this method which lead to excellent results [20].

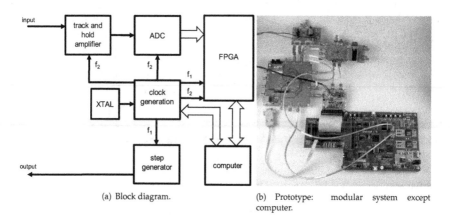

<table>
</table>

(a) Block diagram.   (b) Prototype: modular   system   except
                                            computer.

Figure 4. Time domain transmission measurement system with a FPGA.

mostly in the digital part behind the analog-digital converter (ADC); the front end was very similar.

A block diagram and a photograph of the first system using a microcontroller are shown in Figure 3. Two slightly detuned clocks are synthesized from a crystal. One clock triggers a step generator, which transmits a step signal with about 30ps rise time (20% to 80%) and up to 3V amplitude. The other triggers the track and hold amplifier (having an input bandwidth of 13GHz), the 12bit ADC, and the asynchronous FIFO (first in first out) memory with a width of 18bit and a depth of 32k. A microcontroller and a computer control the system. The FIFO memory can be read out, and the clock generation can be programmed via the I2C-bus. Clock signals up to hundreds of megahertz can be chosen with an accuracy of 1ppm. The dimensions of the RF-PCB are 90mm × 60mm. A reference path is necessary to measure time differences, because the phase of the two clock signals is not captured. A FIFO has to be used because the microcontroller is not able to read in 12bit words at tens of megahertz. Therefore, one waveform of up to 32k is captured at a time and is then transferred to the computer. Averaging, in order to improve the SNR, is carried out on the computer.

An improvement is the application of a field programmable gate array (FPGA). A modular system is shown in Figure 4. There is no more need for a fast external FIFO because data can be read into the FPGA directly up to about 66MHz. Furthermore, no reference path is needed, because the FPGA is able to discriminate the phase between both clock signals. Up to $2^{12} = 4096$ sampling points are possible while an averaging of up to $2^{13} = 8192$ waveforms can be carried out on the FPGA (this is restricted due to the internal resources of the FPGA used[2]).

Since both front ends are similar they offer a comparable performance. The RMS noise level of the receiver is < 0.9mV, the RMS jitter is < 0.7ps (no averaging). Figure 5 shows a comparison of sampled test signals between the proposed hardware and a Tektronix TDS8000. The frequencies are chosen as $f_1 = 50.01$MHz and $f_2 = 50$MHz, which leads to a resolution

---
[2] A Spartan 6 from Xilinx is used.

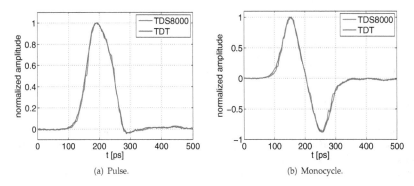

(a) Pulse.   (b) Monocycle.

Figure 5. Comparison between the proposed dedicated hardware (TDT) and a Textronix TDS8000 (no averaging).

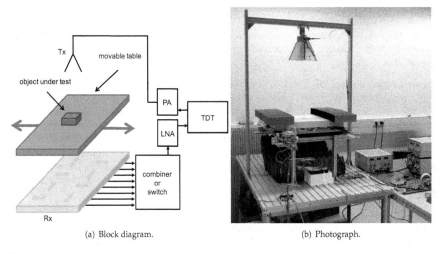

(a) Block diagram.   (b) Photograph.

Figure 6. Entire measurement set-up.

of $\Delta T = 4$ps. For the measurements presented later a much lower resolution is sufficient, e.g. $\Delta T = 40$ps.

3.3. Entire measurement setup

The entire measurement system is shown is Figure 6. The step signal of the TDT is amplified by a power amplifier (PA) before being transmitted through a horn antenna. The object under test is illuminated by the transmitted signal and portions of the scattered signal are received by an array of eight receiving antennas, which are arranged in two orthogonal polarizations. In order to emulate a 2-dimensional array, the movable table is moved orthogonally to the orientation of the receiving array. It is crucial to the multivariate calibration, which is applied later, that the received signals of each antenna placed at its unique position are sampled separately. Therefore, the antennas are switched sequentially onto the input of the TDT and are sampled there, after being amplified by a low noise amplifier (LNA). The MOSFET switch used requires relatively high hard- and software effort and it has to be synchronised with the other components included in the system. Furthermore, these switches have a latency of some nanoseconds. The switch has an insertion loss of 4dB at 2GHz and 8.6dB at 8GHz.

A simpler solution is to use a broadband combiner. The signals received by the individual antennas are combined using a broadband eight-way Wilkinson divider. Prior to combining there is a delay of $\tau = 2ns$ between two adjacent inputs. It is possible to separate the individual pulses in time. Compared to using a receiver having more channels, or using a switch, the hardware effort is greatly reduced and the instantaneous sampling of all pulses is possible in about 20ns. One of the disadvantages that has to be considered is that there is an increased insertion loss of ideally 9dB; at 1GHz and 5GHz the measured insertion losses are 9.3dB and 10.7dB, respectively. There are also losses and temperature dependencies due to the delay lines, which have to be taken into account.

4. Measurements

For the multivariate analysis applied later it is necessary to measure a large number of objects varying in their dielectric and geometric properties. Here, two series of measurements are presented.

The first series comprised objects of moist clay granules. Five irregularly shaped moulds were manufactured from polystyrol foam[3]. They were then filled with clay granules having different amounts of moisture. One of the moulds is shown in Figure 7. The moisture content was varied in a range from about 4.5% to 24%[4] and reference measurements were carried out with a gravimetric method[5]. Furthermore, three rotation angles of $0°$, $22.5°$ and $45°$ were applied to the test objects when putting them on the movable table in order to have more variation in the received signals. Overall 90 different objects were measured. The measurements were carried out in random order to avoid correlations with environmental effects, e.g. temperature variations.

Since there are eight receiving antennas and the movable table is moved to four positions, 32 pulses are received for every object under test when using a switch. With the Wilkison divider and delay lines, the pulses of the eight antennas are received instantaneously as a series of pulses. Two of the received series of pulses for two different objects having moisture contents of 4.88% and 20.93% are shown as examples in Figure 8. Although the variation in moisture content is high, the variation in the pulse shape is rather low.

The second series comprised plastic bottles filled with ethanol-water mixtures[6]. The water content was varied in a range from 2% to 20% in steps around 2%; reference measurements were carried out using a precision balance[7]. Ten bottles were filled with a total of around 190g

---

[3] In the following: *test series 1*

[4] All given moisture contents are on a wet basis.

[5] A Sartorius MA100 is used, accuracy of the weighing function: 0.1% for samples $> 1g$ and 0.02% for samples $> 5g$. The weight of the samples was about $4 - 8g$.

[6] In the following: *test series 2*

[7] The gravimetric water content was determined using a Kern EW 4200-2NM. The repeatability is 0.01g.

(a) One of the moulds manufactured from Styrofoam. (b) Mould filled with moist clay granules. (c) A bottle filled with an ethanol-water mixture.

Figure 7. Objects under test.

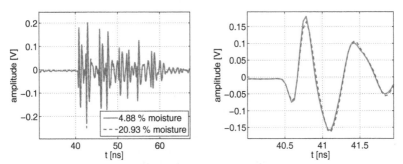

(a) Instantaneous sampled pulses received by the eight antennas using a broadband Wilkinson divider and delay lines.

(b) Closer look at the first received pulse.

Figure 8. Examples of received pulses from two objects having different moisture content.

of liquid, and another ten with about 140g. Therefore, overall 20 objects were measured. A photograph of one of the measured objects containing 113.15g ethanol and 27.94g of water is shown in Figure 7.

The time domain signals are gated in order to extract the time interval of interest: about 10 to 30 equidistant amplitude values per pulse are empirically chosen and subjected to the multivariate calibration methods which will be explained in the next section 5.

5. Multivariate data processing

The sequence of the eight sampled pulses consists of 217 sampling points. There are four positions of the movable table which then leads to $4 \times 217 = 868$ data points per object. These pulses are modified by many factors: the shape, the position, the rotation, and the intrinsic variables of the material under test. However, as can be seen in Figure 8 the value to be measured has only a relatively low influence on the apparent shape of the pulses, and practically all data points are modified when one or more of the factors mentioned above changes. Indeed each factor modifies the curve shape in a different manner. Therefore all measured points of the curve contribute in part to the variable(s) to be measured. Often these values consist only of one variable (for instance water content), a small set of variables (complex permittivity, quality) or an abstract class (shape of the object). Hence the challenge of data processing for the application discussed here is to extract the (hidden) relevant information from a huge data array. Due to the complexity physical modelling is impracticable.

Multivariate calibrations are established techniques for the extraction of relevant information from observed (measured) data without physical modelling. In the following, principal components analysis and regression (PCA/PCR), artificial neural networks (ANN) and partial least squares regression (PLSR) are applied to the data measured during the experiments described in section 4. Multivariate calibration methods have the disadvantage that they require a calibration procedure i.e. training. This means a portion of the measurements carried out on known samples need to be used to determine parameters or coefficients that enable a determination of the variable to be measured for unknown samples.

For this reason the measurements are divided randomly into a calibration and validation group. In general these two groups have an equal size. For test series 1 the number of data sets of the calibration and validation groups is $n_c = n_v = 45$ and for test series 2 they are $n_c = n_v = 10$. The more samples are available the more robust is the calibration and the meaningfulness of the validation.

In order to reduce the amount of data, in a pre-processing step the points having a lower variance may be removed from the input variables. For the experiments described here a majority of the 868 time points can be neglected when the threshold value of the standard deviation is set to 20% of the maximum standard deviation[8]. Thus for test series 1 and 2 the number of points used is $m_1 = 305$ and $m_2 = 187$, respectively. The raw matrix of the calibration data consists of the measured and pre-processed pulse sequence in each line. Hence the columns contain the data of the selected time points of a measurement:

$$
Y_c = \begin{bmatrix} y_{11} & \cdots & y_{1m} \\ \vdots & \ddots & \vdots \\ y_{n_c1} & \cdots & y_{n_cm} \end{bmatrix}. \tag{2}
$$

Due to numerical reasons it is advantageous to standardize the raw data. Firstly the means of each column are calculated with

$$
\overline{Y}_c = \begin{bmatrix} \frac{1}{n_c} \sum_{i=1}^{n_c} y_{i1} & \cdots & \frac{1}{n_c} \sum_{i=1}^{n_c} y_{im} \end{bmatrix}. \tag{3}
$$

The matrix of the normalized, standardized calibration data is calculated by subtracting the means of each column from each value of the columns, then dividing each by the standard deviations of the columns $\sigma_{c1} \cdots \sigma_{cm}$:

$$
X_c = \begin{bmatrix} Y_c - \begin{bmatrix} 1 \\ \vdots \\ 1 \end{bmatrix} \cdot \overline{Y}_c \end{bmatrix} \cdot \begin{bmatrix} \frac{1}{\sigma_{c1}} & \cdots & 0 \\ \vdots & \ddots & \vdots \\ 0 & \cdots & \frac{1}{\sigma_{cm}} \end{bmatrix}. \tag{4}
$$

---

[8] which is 0.0037 and 0.0017 for test series 1 and 2, respectively.

The matrix of the pre-processed validation data is calculated similarly. However the data is normalized and standardized using the means and the standard deviations of the calibration set.

5.1. Principal component analysis and regression

As can be seen in Figure 8 neighbouring data points are highly correlated. Therefore it is not possible to use the selected data points directly in a linear regression to estimate the variable of interest[9]. For the two test series described in section 4 such a linear calibration equation would have $m_1 = 305$ and $m_2 = 187$ coefficients. Furthermore, and this aspect is more relevant, the calculation of the coefficients is numerically unstable because a matrix with correlated data needs to be inverted in the regression algorithm.

A solution to this problem is found using principal component analysis (PCA). The original data is linear transformed into a new set of variables

$$H_c = X_c \cdot P. \tag{5}$$

H_c comprises the so called principal components (the scores) and P is a matrix of the so called loadings. The scores have the advantage that they are uncorrelated and are arranged in such a way that the first principal component has the highest variance and the others are arranged in decreasing order of variance.

The matrix of the loadings P is composed of the eigenvectors. They are also orthogonal and of unit length. This transformation can be interpreted as a transformation into a new orthogonal coordinate system. The basis vectors of the new coordinate system are the eigenvectors and their direction is along the variances in decreasing order.

The properties of the principal components, their orthogonality and their arrangement regarding the variance, enable data reduction because the relevant information of the matrix is already described by the first few principal components. In order to obtain the properties of the transformed data as described above, the matrix P needs to be calculated by an eigenvalue decomposition [36–38] but this is not described in detail here. For the results calculated here the statistical toolbox of MATLAB is used. The eigenvalue decomposition of the PCA is processed without any consideration of the variable(s) of interest. This will be done in the next step of the data processing.

As mentioned above a multiple linear regression of the untransformed (therefore correlated) data is numerically unstable, but after the transformation (see eq. (5)) the data is uncorrelated and the value to be determined can be estimated by a linear combination of the principal components (principal component regression, PCR):

$$\hat{z}_c = \tilde{H}_c \cdot \beta, \tag{6}$$

where $\hat{z}_c$ is the estimated variable of interest (objective variable), $\tilde{H}_c$ is the matrix of the selected principal components and the vector β contains the coefficients of the linear equation.

---

[9] The variable of interest or objective variable is the parameter to be determined later, e.g. the moisture content.

The entries in the first column of $\tilde{H}_c$ are all unity in order to describe the mean of the value of interest in the linear equation.

In $\tilde{H}_c$ only the first k principal components are included. This selection leads to the desired data reduction. The value k need to be determined heuristically. Here for test series 1 the number of principal components used is $k_1 = 12$ and for test series 2 it is $k_2 = 2$.

The coefficients of the linear equation can be calculated by the following equation:

$$\beta = \left(\tilde{H}_c^{\ T} \cdot \tilde{H}_c \right)^{-1} \cdot \tilde{H}_c \cdot z_c, \tag{7}$$

where z_c consists of the variable of interest determined by a reference method, e.g. oven drying for moisture content.

After the calibration data is processed the system is essentially calibrated and is ready to handle unknown samples. However prior to that, the performance of the calibration still needs to be evaluated using the validation data. The target variables of interest are also determined for the pre-processed validation data using a reference method. The validation data (or later in use, the data of a measurement of an unknown sample) is processed in the following manner:

1. the scores are estimated using the loadings determined during the calibration procedure: $\hat{H}_v = X_v \cdot P$,

2. the unused principal components are removed and the unit column is added: $\hat{H}_v \Rightarrow \tilde{\hat{H}}_v$,

3. the value of interest is estimated by the linear equation $\hat{z}_v = \tilde{\hat{H}}_v \cdot \beta$.

For the evaluation of the quality of the calibration the root mean square error of calibration group $RMSE_c$ and the validation group $RMSE_v$ are calculated.

The results obtained with PCA/PCR for both test series are shown in Figure 9. The predicted moisture or water content is plotted vs. its true values. With perfect prediction all points of

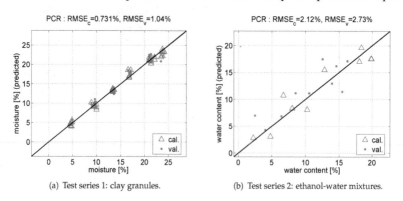

(a) Test series 1: clay granules.   (b) Test series 2: ethanol-water mixtures.

Figure 9. Results obtained with PCA/PCR for both test series.

RER	Classification	Application
Up to 6	Very poor	Not recommended
7-12	Poor	Very rough screening
13-20	Fair	Screening
21-30	Good	Quality control
31-40	Very good	Process control
41+	Excellent	Any application

Table 1. *Classification using RER-values according to [39].*

the calibration and validation group would be on the so called quality line. For test series 1 a $RMSE_c = 0.731\%$ and $RMSE_v = 1.04\%$ is achieved. For test series 2 the errors are $RMSE_c = 2.12\%$ and $RMSE_v = 2.73\%$. The meaningfulness of $RMSE$ depends on the range of the variable to be predicted. Therefore the range error ratio RER is a better choice to evaluate the calibration. It is the ratio between the variable range $\Delta z = \max z - \min z$ and the RMSE[10]:

$$RER = \frac{\Delta z}{RMSE}. \tag{8}$$

The quality of the performance can be assessed using the ranges suggested in Table 1. For test series 1 the $RER_c = 26.7$ and $RER_v = 18.75$, hence the performance is *good*. But for test series 2 the accuracy obtained with PCA/PCR is only *poor* because $RER_c = 8.5$ and $RER_v = 6.6$.

5.2. Artificial neural networks

Although PCA/PCR is a linear operation it is more or less capable of processing non-linear data. However, when the unknown function describing the relationship between the pulse sequence and the value of interest is non-linear a purely linear method may not be the best choice. Artificial neural networks (ANN) can approximate unknown non-linear functions. For this application multilayer-feed-forward (MLFF) networks have a suitable architecture [40].

Such a network is shown in Figure 10. The input variables are weighted and processed by the neurons of the hidden layer. The activation functions of the neurons are non-linear[11]. This enables the non-linear function approximation. The output variable of the hidden layer is weighted again and processed by the neuron(s) of the output layer. The output of this layer needs to be post-processed (scaling and mean value) and the estimated variable of interest is available.

Due to their architecture ANN have several degrees of freedom: the number of hidden layers, the kind of activation function in each layer, and the number of neurons in the hidden layer. For the application discussed here one hidden layer and $n_{HL} = 10$ neurons in this hidden layer are sufficient. The problem is that the number of weighting factors between the layers increases with the number of neurons and for an optimal determination of the weighting factors a relatively large number of samples for the calibration (training) is necessary.

---

[10] In [39] the standard error is used instead of the root mean square error; for large numbers of samples there is practically no difference.

[11] tansig-function

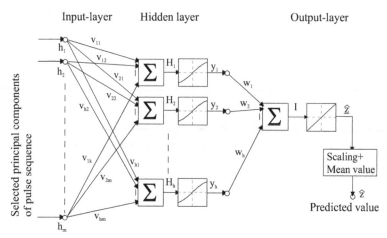

Figure 10. Architecture of the used multilayer-feed-forward-ANN.

For this reason a pre-processing of the data is recommended. If all selected time points were to be fed into the input layer $m_1 n_{HL} = 3050$ and $m_2 n_{HL} = 1870$ weighting factors would need to be found, with only $n_c = 45$ (test series 1) $n_c = 10$ training data sets. Therefore it is useful to feed the ANN with the selected principal components because they include the relevant information. This means the linear principal components regression is replaced by the non-linear ANN.

The training of the ANN has a relatively high calculation effort. Furthermore the starting values for the weighting factors are set randomly at the beginning of the training. This means the method is not strongly deterministic and it is not known for example, whether the optimal weighting factors were found because the training stopped in a local minimum of the error function. The training of the ANN was effected using the artificial neural network toolbox of MATLAB.

The results of the ANN are plotted in Figure 11. In comparison to the results of PCA/PCR there is an improvement observable:

- for test series 1 the RER_c increases to 37.8 (rating: *very good*) and the $RER_v = 22.4$ (*good*). This means there is a slight overfitting,

- for test series 2 the following ratings are obtained: $RER_c = 25.5$ (*good*) and $RER_v = 18.1$ (*fair*).

Despite the much higher calculation effort of ANN the improvements are not very satisfactory.

5.3. Partial least squares regression

PCA decorrelates the data by eigenvalue decomposition. Therefore the variable(s) of interest are not considered in this procedure. Only at the stage of PCR are they taken into account and a selection of relevant principal components is necessary. With the partial-least-squares

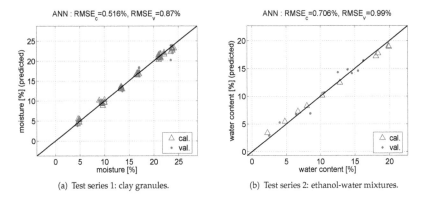

(a) Test series 1: clay granules.   (b) Test series 2: ethanol-water mixtures.

Figure 11. Results obtained with ANN for both test series.

regression (PLSR) the data is decorrelated regarding the variables(s) of interest. Several PLSR algorithms exist and sometimes the data is pre-processed non-linearly. Although PLSR was developed, more or less intuitively, in order to analyze economic data, in the meantime this method has also been used for several applications in other fields.

The algorithm used here for the processing of the measured data is described in [7] in detail and is only summarized in the following.

- Firstly, the input values are weighted in such a way that the covariance between them and the variable of interest is maximal.

- Secondly, the projection of the input values on the vector of the weighting values is called a *factor* or a *hidden path variable*. In the following, two regression analyses are considered:

  1. between the input variables and the factor, and
  2. between the variable of interest and the factor.

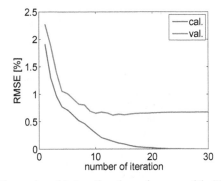

Figure 12. Influence of the number of factors H on the performance of the PLSR.

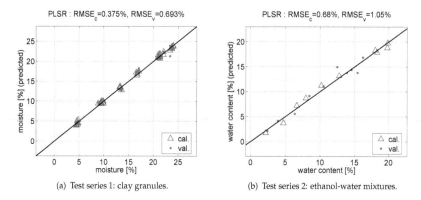

(a) Test series 1: clay granules.   (b) Test series 2: ethanol-water mixtures.

Figure 13. Results obtained with PLSR for both test series.

- Thirdly, the parts described by these linear models are subtracted from the measured values and the variables of interest and the algorithm is restarted using this new data in order to calculate the next factor,

- and finally, the procedure is repeated iteratively until a specified number H of hidden path variables is calculated.

All determined regression coefficients and weighting factors are used finally for the calculation of the regression equation. This means for the validation (and later application) that only a linear combination of the input values need to be calculated. Hence the calculation effort is much smaller in comparison to the ANN. The only degree of freedom is H, the number of factors to be used (number of iterations). When H is too high the $RMSE_c$ is significantly smaller than the $RMSE_v$. This means that overfitting occurs. However, as shown in Figure 12 H should be selected where $RMSE_v$ has a minimum. Furthermore $RMSE_c$ should not be much smaller (factor $1/2$) than $RMSE_v$, otherwise the PLSR calibration could not handle unknown samples.

The performance of the PLSR is shown in Figure 13. For test series 1 $RER_c = 52$ (*excellent*) and $RER_v = 28.1$ (*good*). This is a further improvement in comparison to the ANN. For test series 2 the results stay similar to those of ANN: $RER_c = 26.5$ (*good*) and $RER_v = 17.1$ (*fair*).

5.4. Best calibration method

Principal component analysis and regression lead to acceptable results but the best calibrations were obtained with ANN and PLSR. However the computation effort is much higher with ANN and in general more samples are necessary for a successful training. PLSR is a linear operation and can be performed fast in real time. For this reason PLSR is the best choice for calibration of the application discussed here. In Table 2 the results of both test series are compared with similar experiments presented in several other publications. As can be seen, the performance obtained here is in the upper range. However one has to take into account the further advantages of the system discussed here: it is non-contacting, the objects can be rotated, and can have irregular shapes and sizes.

Publication	range [%]	$RMSE_c$ [%]	$RMSE_v$ [%]	RER_c	RER_v
[41]: Tobacco, PLSR	10-50	-	2	-	20
[42]: Scots pine, PLSR	0-15	0.46	0.74	32.6	20.3
[42]: Scots pine, PLSR	0-175	15.92	12.52	11	14
[9]: Clay granules, ANN	6.3-34.2	1.6	2.1	17.4	13.3
ISOperm:					
Clay granules, PLSR	4.5-24	0.38	0.69	52	28.1
Ethanol water mix. in bottle, ANN	2-20	0.71	1	25.5	18.1
[14], PLSR	70-100	1.28	2.55	23.4	11.8
[16], ANN	5-29.2	1.29	1.88	18.8	12.9
[17], PLSR	4.6-24.1	0.35	0.69	55.3	28.2
[19], PLSR	1.8-20.2	0.39	0.61	47.3	30.2
[20], PLSR	4.3-23.4	0.31	0.55	61.7	34.7
Other technologies:					
[43]: Wheat, admittance, PCR	9-20	-	0.39	-	28.2
[44]: Salmon, NIR, PLSR	61-70.8	-	0.98	-	10
[45]: Paper, NIR, PLSR	0-2.4	-	0.056	-	43.1
[46]: Theophyllin, NIR, ANN	1-22	0.45	0.83	46.7	25.3

Table 2. Comparison to other publications regarding the accuracy of the determination of moisture or water content. Except the method investigated in ISOPerm, all others are contacting and/or require a defined shape of the object under test.

6. Conclusions

Many industrial and scientific applications require extensive on-line process monitoring and quality control. Often the composition of goods (e.g. moisture content) is of great interest but also abstract parameters, for example quality or freshness, play an important role. The microwave sensor described is able to penetrate the investigated materials and by using UWB-techniques it is possible to gain information at various frequencies. The applied time domain techniques operate with low hardware effort and fast measurement speed while having a high accuracy. Using commercial MMICs signals exceeding a bandwidth of 10GHz can be generated and sampled with cheap and compact dedicated hardware. Today it is possible to employ multivariate calibration methods like artificial neural networks, which have a high computational effort, in real time. These methods are well established in, for example, NIR or image processing and are successfully adopted. The feasibility of the method has been successfully proven with accuracy even greater than in many previous publications using contacting methods. It has a great potential for many kinds of future applications in microwave sensing.

Author details

Henning Mextorf, Frank Daschner, Mike Kent and Reinhard Knöchel
University of Kiel, Germany

7. References

[1] Vainikainen, P. & Laitinen, T. [2009]. *Proceedings of the 8th International Conference on Electromagnetic Wave Interaction with Water and Moist Substances (ISEMA 2009)*, Espoo, Finland.

[2] Grant, E. H., Sheppard, R. J. & South, G. P. [1978]. *Dielectric behaviour of biological molecules in solution*, Clarendon Press, Oxford, UK.

[3] Fellner-Feldegg, H. [1969]. The measurement of dielectrics in the time domain, *J. Phys. Chem.* 73: 616 – 623.

[4] Burdette, E., Cain, F. & Seals, J. [1980]. In vivo probe measurement technique for determining dielectric properties at VHF through microwave frequencies, *IEEE Trans. Microw. Theory Tech.* 28(4).

[5] Kent, M. & Anderson, D. [1996]. Dielectric studies of added water in poultry meat and scallops, *Journal of food engineering* 28(3-4): 239–259.

[6] Kent, M., Knöchel, R., Daschner, F. & Berger, U.-K. [2000]. Composition of foods using microwave dielectric spectra, *Eur. Food Res. Technol.* (210): 359–366.

[7] Martens, H. & Naes, T. [1989]. *Multivariate Calibration*, John Wiley and Sons, Chichester.

[8] Kent, M., Knöchel, R., Daschner, F., Schimmer, O., Oelenschläger, J., Mierke-Klemeyer, S., Barr, U.-K., Floberg, P., Huidobro, M., Nunes, L., Batista, I. & Martins, A. [2004]. Time domain reflectometry as a tool for the estimation of quality in foods, *Agrophysics* 18(3).

[9] Schimmer, O., Gülck, A., Daschner, F., Piotrowski, J. & Knöchel, R. [2005]. Non-contacting determination of moisture content in bulk materials using sub-nanosecond UWB-pulses, *IEEE Trans. Microw. Theory Tech.* 53(6): 2107–2113.

[10] Mextorf, H., Martens, R., Daschner, F. & Knöchel, R. [2010]. Dual polarized UWB antenna for free-space characterization of dielectric objects, *Proc. German Microwave Conference 2010* pp. 162 – 165.

[11] Mextorf, H., Daschner, F., Kent, M. & Knöchel, R. [2010a]. Free-space determination of permittivity, size and orientation of rectangular shaped objects using multivariate analysis, *Proc. European Microwave Conference 2010* pp. 152 – 155.

[12] Mextorf, H., Daschner, F., Kent, M. & Knöchel, R. [2010b]. Non-contacting UWB-characterization of dielectric objects using multivariate calibration, *Proc. Aquametry 2010* pp. 136 – 144.

[13] Mextorf, H., Daschner, F., Kent, M. & Knöchel, R. [2011d]. UWB free-space characterization and shape recognition of dielectric objects using statistical methods, *IEEE Trans. Instrum. Meas.* 60(4): 1389 – 1396.

[14] Mextorf, H., Daschner, F., Kent, M. & Knöchel, R. [2011a]. Free-space prediction of the water content of irregularly shaped bodies filled with water-ethanol mixtures, *Proc. ISEMA 2011* pp. 162 – 169.

[15] Mextorf, H., Daschner, F., Kent, M. & Knöchel, R. [2011b]. New UWB free-space method for the classification and characterization of dielectric objects, *Proc. ICUWB 2011* pp. 410–414.

[16] Mextorf, H., Daschner, F., Kent, M. & Knöchel, R. [2011c]. Performance of multivariate calibration methods for the UWB characterization of dielectric objects, *Proc. CMM-Tagung 2011* pp. 105–112.

[17] Mextorf, H., Daschner, F., Kent, M. & Knöchel, R. [2012a]. Non-contacting moisture sensing using a dedicated UWB time domain instrument, *Proc. German Microwave Conference 2012* pp. 1–4.

[18] Mextorf, H., Daschner, F., Kent, M. & Knöchel, R. [2012b]. Signal quality considerations for free-space UWB moisture measurements, *Proc. Mikon 2012* pp. 627–630.

[19] Mextorf, H., Daschner, F., Kent, M. & Knöchel, R. [2012c]. UWB time domain transmission sensor for free-space moisture measurements, *IEEE MTT-S Int. Microw. Symp. Dig. 2012* pp. 1–3.

[20] Mextorf, H., Sachs, J., Daschner, F., Kent, M. & Knöchel, R. [2012]. Free-space moisture prediction of small objects using M-sequences, *Proc. ICUWB 2012* pp. 260–264.

[21] Schantz, H. G. [2003a]. Introduction to ultra-wideband antennas, *IEEE Conference on Ultra Wideband Systems and Technologies* pp. 1–9.

[22] Schantz, H. G. [2003b]. UWB magnetic antennas, *IEEE Antennas and Propagation Society International Symposium* 3: 604–607.

[23] Schantz, H. G. [2004]. A brief history of UWB antennas, *IEEE Aerospace and Electronic Systems Magazine* 19(4): 22–26.

[24] Schantz, H. G. [2005]. *The Art and Science of Ultrawideband Antennas*, Artech House, Norwood, MA.

[25] Wiesbeck, W., Adamiuk, G. & Sturm, C. [2009]. Basic properties and design principles of UWB antennas , *IProceedings of the IEEE* 97(2): 372–385.

[26] Perruisseau-Carrier, J., Hee, T. W. & Hall, P. S. [2003]. Dual-polarized broadband dipole, *IEEE Antennas Wireless Propag. Lett.* 2(1): 310–312.

[27] Woten, D. A. & El-Shenawee, M. [2008]. Broadband dual linear polarized antenna for statistical detection of breast cancer, *IEEE Trans. Antennas Propag.* 56(11): 3576–3590.

[28] Teo, P.-T., Lee, K.-S. & Lee, C.-K. [2003]. Maltese-cross coaxial balun-fed antenna for GPS and DCS1800 mobile communication, *IEEE Trans. Veh. Technol.* 52(4): 779–783.

[29] Suh, S.-Y., Stutzman, W., Davis, W., Walthot, A. & Schiffer, J. [2004]. A generalized crossed dipole antenna, the fourtear antenna, *IEEE APS 2004* 3: 2915–2918.

[30] Mak, K.-M., Wong, H. & Luk, K.-M. [2007]. A shorted bowtie patch antenna with a cross dipole for dual polarization, *IEEE Antennas Wireless Propag. Lett.* 6: 126–129.

[31] Adamiuk, G., Beer, S., Wiesbeck, W. & Zwick, T. [2009]. Dual-orthogonal polarized antenna for UWB-IR technology, *IEEE Antennas Wireless Propag. Lett.* 8: 981–984.

[32] Soroka, S. [1986]. A physically compact quad ridge horn design, *IEEE APS 1986* pp. 903–906.

[33] Shen, Z. & Feng, C. [2005]. A new dual-polarized broadband horn antenna, *IEEE Antennas and Wireless Propagation Letters* 4: 270–273.

[34] Schaubert, D., Elsallal, W., Kasturi, S., Boryssenko, A., Vouvakis, M. N. & Paraschos, G. [2008]. Wide bandwidth arrays of vivaldi antennas, *Institution of Engineering and Technology Seminar on Wideband, Multiband Antennas and Arrays for Defence or Civil Applications* pp. 1–20.

[35] Adamiuk, G., Zwick, T. & Wiesbeck, W. [2007]. Dual-orthogonal polarized vivaldi antenna for ultra wideband applications, *MIKON 2008* pp. 1–4.

[36] Götze, J. [1989]. *Orthogonale Matrizentransformationen*, R. Oldenbourg Verlag, München.

[37] G. Engeln-Muellges, F. R. [1996]. *Numerik Algorithmen*, VDI-Verlag, Düsseldorf.

[38] G.H. Golub, v. L. [1996]. *Matrix Computations*, The John Hopkins University Press, Baltimore.

[39] Williams, P. C. [2001]. *Near-Infrared Technology in the Agriculture and Food industries*, 2nd edn, American Association of Cereal Chemists, Inc., chapter Implementation of near-infrared technology, p. 165.

[40] Patterson, D. [1997]. *Künstliche neuronale Netze*, Prentice Hall Verlag, Haar.

[41] Dane, A. D., Rea, G. J., Walmsley, A. D. & Haswell, S. J. [2001]. The determination of moisture in tobacco by guided microwave spectroscopy and multivariate calibration, *Analytica Chimica Acta* 429(2): 185–194.

[42] Johansson, J., Hagmana, O. & Oja, J. [2003]. Predicting moisture content and density of Scots pine by microwave scanning of sawn timber, *Computers and Electronics in Agriculture* 41(1-3): 85–90.

[43] Lawrence, K. C., Windham, W. R. & Nelson, S. O. [1998]. Wheat moisture determination by 1- to 110-MHz swept-frequency admittance measurements, *Transactions of the ASAE* 41(1): 135–142.

[44] Wold, J. P. & Isaksson, T. [1997]. Non-destructive determination of fat and moisture in whole atlantic salmon by near-infrared diffuse spectroscopy, *Journal of Food Science* 62(4): 734–736.

[45] Neimanis, R., Lennholm, H. & Eriksson, R. [1999]. Determination of moisture content in impregnated paper using near infrared spectroscopy, *1999 Annual Report Conference on Electrical Insulation and Dielectric Phenomena* 1: 162–165.

[46] Rantanen, J., Rasanen, E., Antikainen, O., Mannermaa, J.-P. & Yliruusi, J. [2001]. In-line moisture measurement during granulation with a four-wavelength near-infrared sensor: an evaluation of process-related variables and a development of non-linear calibration model, *Chemometrics and Intelligent Laboratory Systems* 56(1): 51–58.

Chapter 2

Concepts and Components for Pulsed Angle Modulated Ultra Wideband Communication and Radar Systems

Alexander Esswein, Robert Weigel, Christian Carlowitz and Martin Vossiek

Additional information is available at the end of the chapter

1. Introduction

Ultra Wideband (UWB) systems have been utilized and commercialized since the beginning of the 1970s and have been successfully used in ground-, wall- and foliage-penetration, collision warning and avoidance, fluid level detection, intruder detection and vehicle radar and also for the topics of the intended research project, communication and position-location [1]. Up to now, the latter two fields have been treated separately in most developments.

UWB has the potential to yield solutions for the challenging problem of time dispersion caused by multipath propagation in indoor channels. For a local positioning system, multipath propagation determines the physical limit of the maximal accuracy that can be obtained at a given signal bandwidth [14].

There exist several techniques which are used to generate ultra wideband signals. Traditionally, UWB was defined as pulse based radio. Especially for radar and localization applications, the use of very narrow pulses is still the most dominant technique. In addition to that, there are UWB systems that use more complex modulation techniques, like multiband orthogonal frequency-division multiplexing (MB-OFDM) or direct sequence code-division multiple access (DS-CDMA) to spread the transmitted information over a large bandwidth. They are applied in communication systems whereas radar systems that use such techniques can hardly be found.

Recently there can be recognized an increasing interest for UWB technologies applied in mm-wave frequency bands. This interest is stimulated by novel regulation for future vehicular UWB systems in the 79 GHz band (77 - 81 GHz) [12], novel international allocation of unlicensed bands ranging from 57 - 66 GHz [9] and the attractive ISM bands at 122.5 GHz with 1 GHz bandwidth and at 244 GHz with 2 GHz bandwidth. Also, the 61.5 GHz ISM band with 500 MHz available bandwidth is often considered as a "de-facto" UWB band even though the bandwidth is just less than the bandwidth of 500 MHz usually demanded as the minimal bandwidth for UWB. The great advantage of mm-wave UWB bands is that they do not suffer from the severe power regulations known from standard UWB. At the above mentioned mm-wave UWB bands, the permitted maximum mean power density is at least 38 dB higher than in the UWB bands below 30 GHz.

Most of the mm-wave UWB communication and ranging systems published so far use a simple pulse generator as signal source. In the simplest case, a mm-wave CW carrier is modulated with an ASK (s. e.g [17]) or BPSK (s. e.g. [18])) sequence. A very interesting low-power approach that is somewhat related to the approach in this work is shown in [6] and [7]. Here, a 60 GHz oscillator itself is switched on and off. To guarantee a stable startup phase and to improve the phase noise, the oscillator is injection locked to a spurious harmonic of the switching signal. The benefit of the pulsed injection locking approach with respect to power consumption was impressively shown in this work. The general approach to obtain a stable pulse to pulse phase condition by injecting a spurious harmonic of the switching pulse into the oscillator is well known for a long time from low-power and low-cost microwave primary pulse radar systems. This basic principle can be extended in a way that frequency modulated signals can be generated based on a switched injection locked oscillator [19]. In this work, it is generalized for synthesizing arbitrarily phase modulated signals for integrated local positioning and communication. The fusion of positioning and communication capability is especially needed for future wireless devices applied in the "internet of things" or for advanced multimedia / augmented reality applications, for robot control and for vehicle2X / car2X applications.

Most existing UWB communication and ranging systems - especially those dedicated to low power consumption and mm-wave frequencies - employ simple impulse radios (IR). Popular IR-UWB modulation techniques include on-off keying (OOK), pulse-position modulation (PPM), pulse-amplitude modulation (PAM) and binary phase shift keying (BPSK) [5, 17, 18]. Their waveform can be synthesized using low complexity impulse generators and control circuitry, which comes at the cost of low spectral efficiency and severely limited control over spectral properties of the synthesized signals. Consequently, these transmitter cannot exhaust regulatory boundaries in all operation modes. High data rate synthesizers are often average power limited whereas low data rate implementations may be peak power limited [20].

2. Proposed concepts and components

In order to overcome these issues, pulsed angle modulated UWB signals are proposed to provide greater flexibility and better control over the spectral properties of the synthesized signals. Additionally, this signal type is well suited for both ranging and communication, since it allows synthesizing pulsed frequency modulated chirps that are attractive for ranging as well as digital phase modulation schemes for data transmission with the same hardware.

Since classic architectures containing VCOs, PLLs, mixer, linear amplifiers and switches are not suited for low complexity, low power systems, the switched injection-locked oscillator is suggested for signal synthesis. It regenerates and amplifies a weak phase-modulated signal. Consequently, the high frequency RF signal can be generated from a high power but efficiently synthesized low frequency phase modulated baseband signal in two simple stages - a lossy passive or low power frequency multiplier (harmonic generator) and a switched injection-locked oscillator as single stage pulsed high gain (> 50 dB) amplifier.

In this work, it is demonstrated that this approach allows synthesizing pulsed, arbitrarily phase modulated signals using the switched injection-locked harmonic sampling principle. The theory of this concept was investigated thoroughly and verified experimentally for the synthesis of phase shift keying (PSK) modulated communication signals and pulsed frequency modulated (PFM) radar signals with the same hardware. Regarding the switched injection-locked oscillator, implementations in planar surface mounted technology (6-7, 7-8 GHz) and integrated circuits (6-8 GHz, 63 GHz) were developed. Measurements with the first designs confirm the feasibility of the proposed concepts and already show promising results regarding transmitter signal to spur ratio and achievable ranging resolution and ranging uncertainty.

This work shows the half-term results of the ongoing project "Components and concepts for low-power mm-wave pulsed angle modulated ultra wideband communication and ranging (PAMUCOR)" within the DFG priority programme "Ultra-Wideband Radio Technologies for Communications, Localization and Sensor Applications"; for comparison, some results from the previous project "Concepts and components for pulsed frequency modulated ultra wideband secondary radar systems (PFM-USR)" are summarized.

3. Pulsed angle modulated UWB signals

3.1. Signal definition

Fig. 1 depicts a pulsed angle modulated UWB signal consisting of a sequence of short pulses (width T_d, period T_s), in which each pulse is an oscillation with the frequency ω_{osc} and the modulated initial phase φ_i:

$$s(t) = \sum_{i=0}^{N} \cos\left(\omega_{osc}\left(t - i \cdot T_s + \frac{T_d}{2}\right) + \varphi_i\right) \cdot \text{rect}\left(\frac{t - i \cdot T_s}{T_d}\right) \tag{1}$$

with

$$\text{rect}(x) = \begin{cases} 1 & \text{for } -0.5 < x < 0.5 \\ 0 & \text{else} \end{cases}.$$

For flexible signal synthesis, initial phase modulation, pulse period, pulse width and oscillation frequency can be tuned.

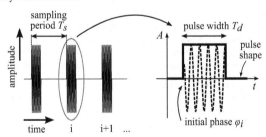

Figure 1. Pulsed angle modulated UWB signal - the modulated parameter is the initial phase φ_i of each pulse

3.2. SILO operation principle

The switched injection-locked oscillator (SILO) is basically a normal oscillator which is turned on and off while a weak reference signal is injected into its feedback loop (see Fig. 2). During startup of the oscillator, the injection signal provides an initial condition in the oscillator's resonator instead of noise like in oscillators without injection signal. This way, the instantaneous phase of the injection signal is adopted though the oscillator runs with its own natural frequency, which may differ from the injection signal's frequency. Since the power level of the injection signal is far too low to influence the oscillation as soon as the oscillator has reached its final amplitude, it performs only phase, but no frequency locking.

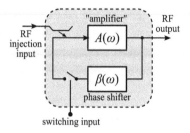

Figure 2. SILO principle

This behavior can be described theoretically by:

$$s(t) = \sum_{i=0}^{N} \cos\left(\omega_{osc}\left(t - i \cdot T_s + \frac{T_d}{2}\right) + \arg\left\{s_{inj}(t)\right\}\right) \cdot \mathrm{rect}\left(\frac{t - i \cdot T_s}{T_d}\right), \qquad (2)$$

with the injection signal (center/reference frequency ω_{inj}, phase modulation $\varphi(t)$)

$$s_{inj}(t) = \cos\left(\omega_{inj}t + \varphi_{inj}(t)\right), \quad \arg\left\{s_{inj}(t)\right\} = \omega_{inj}t + \varphi_{inj}(t). \qquad (3)$$

In spite of the fact that this model only describes the fundamental principle, the physical behavior of the oscillator is very similar in most operation modes. The most important disregarded physical effects observed in real implementations are:

- Due to balancing imperfections e.g. in differential oscillators, high order harmonics of the startup pulse turning on the circuit cause self-locking effects that degrade the SILO's performance at low injection levels. Hence, the rise time of the oscillator should not be too short in order to reduce the harmonic power level. Obviously, this leads to a trade-off with spectral bandwidth, minimum pulse width and maximum achievable pulse repetition rate.

- The phase sampling process is affected by the amplitude of the injection signal. In consequence, amplitude variations of the injection signal are converted into phase distortions. Therefore, constant amplitude injection signals should be used to mitigate these effects. Then there is only a constant phase offset between injection and regenerated signal.

- If the rise time of the oscillator is configured to be relatively long compared to the pulse width, there will be a noticeable dependence between the injection signal's power level and pulse width. With a large amplitude injection signal, the oscillator settles much faster than when starting from noise level. Again, constant amplitude injection signals are the preferred countermeasure to avoid pulse width jitter.

Thus, the simplifications of the proposed ideal model mainly affect time and frequency domain amplitude shape, which makes this model suitable for the analysis of the phase sampling process.

3.3. Phase sampling theory

In [3, 4, 19], the SILO's phase sampling principle and its applications have been investigated thoroughly. The most important results will be summarized and discussed in the following.

Starting from equations (2) and (3), the SILO's output signal can be expressed by (disregarding negative frequencies and finite time domain waveform length for sake of simplicity):

$$s(t) = \sum_{i=-\infty}^{+\infty} \left[a \cdot e^{j\left(\omega_{osc}t + (\omega_{inj}-\omega_{osc})\cdot\left(i\cdot T_s - \frac{T_d}{2}\right)\right)} \cdot e^{j\varphi_{inj}\left(i\cdot T_s - \frac{T_d}{2}\right)} \cdot \text{rect}\left(\frac{t - i\cdot T_s}{T_d}\right) \right]. \tag{4}$$

This expression still suggests an oscillation with ω_{osc} - the presence of the injection signal regeneration feature that includes the frequency is not obvious. According to [4], the Fourier transform $F\{\cdot\}$ of (4) leads to:

$$S(\omega) = A \cdot \left[\text{sinc}\left(\frac{(\omega - \omega_{osc})\cdot T_d}{2}\right) \cdot \left(e^{j(\omega_{inj}-\omega_{osc})\frac{T_d}{2}}\right.\right.$$
$$\left.\left.\cdot F\{e^{j\varphi_{inj}\left(t-\frac{T_d}{2}\right)}\}(\omega) * \delta(\omega - \omega_{inj}) * \text{III}_{\frac{1}{T_s}}\left(\frac{\omega}{2\pi}\right)\right)\right]. \tag{5}$$

The SILO output spectrum according to (5) consists of a convolution of the user-defined phase modulation spectrum with its center / carrier frequency signal and the sampling process' aliasing signal (Dirac comb, III), see Fig. 3. It is weighted with a sinc envelope centered at the oscillator's natural frequency ω_{osc}. Since this frequency only affects the envelope and a constant phase offset, the SILO can be regarded as a highly effective aliased regenerative amplifier. In consequence, an injected user-defined constant envelope phase modulated signal is reproduced correctly even with a free running oscillator with (in certain bounds) unknown natural frequency as long as Nyquist's sampling theorem is fulfilled (modulation bandwidth less than half pulse repetition frequency).

In general, this signal synthesis principle is not limited to phase modulated / constant envelope signal synthesis. For amplitude modulation, e.g. an electronically tuned attenuator at the SILO's output can be employed to manipulate the amplitude of each pulse synchronously to the pulse rate, which leads to a polar modulator. Since efficient pulse

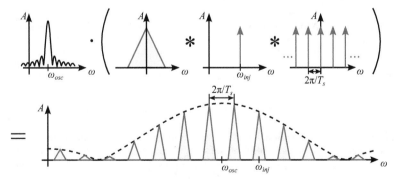

Figure 3. SILO output spectrum according to (5)

amplitude modulation is feasible for a long time in contrast to complex phase modulation and can be added independently, this work is concentrated on the latter aspect.

3.4. Phase modulated UWB communication signals

For the synthesis of communication signals [4], any phase modulated constant envelope signal that is bandwidth limited to half pulse repetition frequency can be chosen. The maximum possible symbol rate leads to one symbol per pulse.

Demodulation can be achieved similar to existing approaches that allow quadrature pulse demodulation (e.g. [11]). Basically, the phase of each pulse has to be sampled synchronously to the pulse sequence (i.e. during pulse duration), which can be realized e.g. by quadrature baseband down-conversion and synchronized sample acquisition. In this case, the sequence of received samples is given by

$$s_{recv}(k) = s\left(k \cdot T_s + \Delta t_{sync}\right) \cdot e^{-j\omega_{inj}\left(k \cdot T_s + \Delta t_{sync}\right)}, \quad k \in \mathbb{N}, \tag{6}$$

where Δt_{sync} denotes a modestly (uncertainty less than half pulse width) unknown synchronization error that has to be taken into account in practice. Inserting (4) in (6) leads to:

$$s_{recv}(k) = A_r \cdot e^{j\left(\varphi_{inj}\left(i \cdot T_s - \frac{T_d}{2}\right) + \left(\omega_{osc} - \omega_{inj}\right) \cdot \left(\Delta t_{sync} + \frac{T_d}{2}\right)\right)}. \tag{7}$$

Accordingly, the original phase modulation φ_{inj} is reconstructed correctly aside from a constant phase offset. Its constancy is guaranteed as long as the natural frequency of the unstabilized oscillator does not drift too fast, which is mostly given due to relatively slow changes in environmental parameters like temperature. For compensation, e.g. differential modulation schemes or short frames can be applied.

3.5. Frequency modulated UWB radar signals

Since the SILO based synthesizer is capable of generating any constant envelope phase modulated signals (within the bandwidth limit), even a frequency modulated radar signal with the bandwidth B, sweep duration T and phase

$$\varphi_{inj,FM}(t) = 2\pi \frac{B}{2T} t^2 \tag{8}$$

can be transmitted. At the receiver, the time delayed transmit signal $s(t)$ is mixed with a FMCW signal:

$$s_{recv,FM}(t) = s(t - t_d) \cdot e^{-j\left(\omega_{inj}t + \pi \frac{B}{T}t^2\right)}. \tag{9}$$

According to [3], the approximate resulting beat frequency spectrum (disregarding envelope)

$$S_{recv,FM}(\omega) = \underline{A} \cdot \delta\left(\omega + 2\pi \frac{B}{T}\left(t_d + \frac{T_d}{2}\right)\right) * III_{\frac{1}{T_s}}(\omega) \tag{10}$$

is equivalent to the conventional FMCW spectrum except for the aliases resulting from switched operation and a constant phase offset $\underline{A}$. The (one way) distance can be calculated from

$$f_b = \frac{B}{T}\left(t_d + \frac{T_d}{2}\right) \tag{11}$$

given that transmitter and receiver were precisely synchronized, which can be achieved through two-way synchronization like in [16].

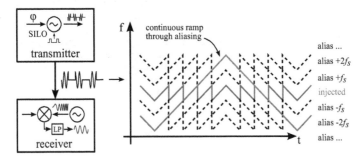

Figure 4. Exploiting sampling aliases to synthesize frequency modulated UWB radar signals with minimal effort

Strictly speaking, the sampling theorem is not met for a sweep bandwidth larger than the pulse repetition frequency. Though, aliasing can be exploited to minimize the ramp synthesis effort (see Fig. 4). The injected and regenerated signal is configured to represent a short chirp within the sampling bandwidth that is repeated continuously. Considering aliasing, the resulting signal appears to be continuous at the receiver when sweeping through all aliases.

The required effort can even be further reduced: Since the SILO only samples certain phase values, it is not necessary to actually generate continuous sweeps as intermediate signal. Instead, a CW injection signal with stepped phase modulation is sufficient as long as its phase (modulus 2π) equals (8) at sampling time. This approach results according to [3] in a short periodic sequence of samples (period $p \in \mathbb{N}^+$) under the condition that the term

$$\frac{pBT_s^2}{T} \tag{12}$$

is whole-number and p even. The sequence features a minimum period of

$$p_{min} = \frac{T}{BT_s^2}, \quad p_{min} \in \mathbb{N}^+. \tag{13}$$

The only restriction that results from exploiting aliases is a limitation in unambiguous range, i.e. maximum distance (phase velocity c_p):

$$d_{max} = \frac{c_p T}{BT_s}. \tag{14}$$

Considering a sampling period of 100 ns ($T_s = 10\,\text{MHz}$), which is convenient for low power implementations, a sufficient maximum range of over 1 km can be achieved even at a high bandwidth of 2 GHz in 1 ms.

4. System concepts

In the following, concepts and implementations for the pulsed angle modulated signal synthesis principle are presented. Firstly, the harmonic sampling approach is presented, which is used to take advantage of all benefits of the switched injection-locked oscillator concept by generating a high power, high frequency signal efficiently from a low frequency intermediate signal (4.1). Secondly, a frequency modulated direct digital synthesis (DDS) based upconversion approach for radar applications from the preceding project (PFM-USR) is presented as starting point for the subsequent development (4.2). Thirdly, the recent hardware concept and implementation for phase stepped modulation is described, which allows for synthesizing both frequency modulated radar signals and phase modulated communication signals with the same simple communication signal generator hardware for integrated communication and ranging.

4.1. Harmonic sampling approach

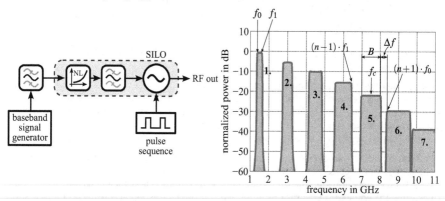

Figure 5. SILO based harmonic sampling; left: concept, right: spectrum of bandwidth limited signal after harmonic generator (here: FMCW sweep from f_0 to f_1) [2]

When synthesizing a high frequency pulsed angle modulated signal, classic approaches based on VCO, PLL, linear amplifier and pulsed switch are not suitable to meet goals like low complexity and low power hardware. Instead, a baseband modulator is proposed for signal generation that generates much lower frequencies than at the system's RF output, e.g. 5.8 GHz instead of 63.8 GHz. At lower frequency ranges, analog RF circuits are usually more efficient than their high frequency counterparts. The baseband signal is then applied to the input of a passive or low power non-linear element that generates harmonics, e.g. a diode or transistor (see Fig. 5). Finally, a SILO is used to amplify the upconverted signal by typically more than 50 dB (within pulse duration). Considering an instantaneous output power level of 0 to 5 dBm, an injection level of less than −45 dBm is sufficient, which allows for high losses and low power consumption in the preceding frequency multiplier stage.

In order to avoid strong intermodulation products caused by the baseband modulation, it should be "slow" compared to the center frequency of the baseband signal so that the non-linear element's instantaneous input and filtered output signal can be considered approximately single tone. This requirement is needed for the SILO, which can itself only correctly regenerate constant envelope signals (apart from the fact that intermodulation products are undesirable) that are stable during the startup phase of the oscillator, e.g. FMCW signals with low ramp slope or rectangular shaped PSK with symbol rate / pulse repetition frequency much smaller than RF frequency.

Regarding maximum baseband modulation bandwidth, there exists a limit for the frequency multiplication factor n in order to guarantee spectral separation, since the bandwidth increases with the harmonic order whereas the spacing of the harmonics' center frequencies is equidistant. According to [2] (see also Fig. 5 right), the upper boundary for the multiplication factor is (harmonic center frequency f_c, harmonic modulation bandwidth B):

$$n < \frac{f_c}{B} - \frac{1}{2}. \tag{15}$$

4.2. Frequency modulated baseband upconversion

The "classic" approach towards synthesizing linear frequency modulated signals (see Fig. 6) consists of a DDS generating a low frequency reference chirp, a PLL and VCO loop and a linear power amplifier. By adding a pulsed switch at the output, pulsed frequency modulation can be realized similar to section 4.2 as long as the pulse width is short enough (the latter signal has constant phase during the pulse, the first one features slight frequency modulation). Obviously, this classic approach has several disadvantages at high frequencies, especially power consuming linear amplifiers and a switch that dissipates more than 90% of the RF power at common pulse sequence duty cycles of less than 1:10.

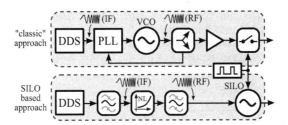

Figure 6. Comparison of classic and SILO based pulsed frequency modulated signal synthesis [2]

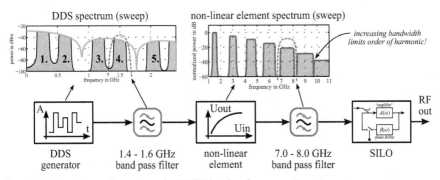

Figure 7. Harmonic sampling concept for FMCW baseband upconversion

Therefore, a harmonic sampling approach was proposed to directly synthesize the ramp from a DDS signal while avoiding PLLs and linear amplifiers at high frequencies [2]. Due to the bandwidth restrictions with harmonic sampling (see section 4.1), a single non-linear stage is not sufficient to generate a 7-8 GHz ramp with a commercially available 1 GS/s DDS circuit. Hence, a Nyquist image from the DDS is used to shift the baseband output frequency range to 1.4-1.6 GHz (see Fig. 7).

The main advantage of this concept is that the generated pulsed frequency modulated signal features a very good linearity in comparison to simple PLL control loops and that the only active component at output frequency is a simple, efficient oscillator (SILO). Despite the simplicity of this concept, its hardware design is quite challenging, since the amplitude of a wideband sweep is subject to many inherent sources of frequency dependent amplitude behavior like DDS spectral envelope, insufficient filter flatness and the non-linear element, which increases existing amplitude variations notably.

4.3. Phase stepped modulation with CW baseband for integrating radar and communication

For integrated communication and ranging, it is desirable to construct a hardware that can synthesize signals for both domains. In the past, they have mostly been developed separately with different hardware concepts. The previously proposed concept (4.2) is well suited for radar systems, but very specific to frequency modulated signals. In fact, angle modulated communication signals can be synthesized with further reduced effort (see Fig. 8) from a CW source with a phase shifter. It is synchronized with the SILO's pulse sequence and its offset is configured to guarantee that each new phase state is stable when the oscillator is turned on. This kind of modulation technique can also be employed to generate frequency modulated signals efficiently according to section 3.5 by using an appropriate sequence of phase samples that represent a frequency chirp.

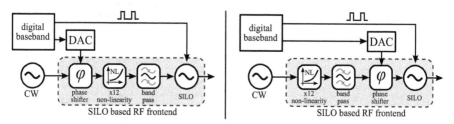

Figure 8. Concept for synthesizing pulsed angle modulated signals using a phase shifter; left: baseband phase modulation with 30 deg shifter, right: RF phase modulation with 360 deg shifter

Regarding hardware implementations, there are two major alternatives concerning the location of the phase shifter in the signal path. An attractive option is to add phase modulation before the frequency multiplication stage; this leads to a minimum amount of RF components and the phase shifter only needs to cover a shifting range of 30 degree, which is easy to design with good linearity. However, baseband modulation limits the multiplication factor (see section 4.1) and the phase shifter causes amplitude fluctuations that are increased in the subsequent non-linear stage. Alternatively, the phase shifter can be placed between RF filter and SILO, which allows for fast modulation, high multiplication factors with less effort (only a constant frequency single tone signal is applied), but requires a more sophisticated 360 degree phase shifter at RF frequency.

5. SILO concept and implementation

Consider the signal displayed in Fig. 1 and the basic SILO model depicted in Fig. 2. As a pulse width T_d of 1 ns and shorter was to be accomplished, the large parasitic capacitances associated with discrete components made it clear that only an integrated solution would be suitable for implementation of the SILO.

As a benchmark for the novel circuit concept of the SILO, some key components of a more conservative concept of generating pulsed frequency modulated signals were developed in an integrated circuit.

All integrated circuits were designed in Cadence Virtuoso and simulated using the Cadence Virtuoso Spectre Circuit Simulator (Cadence, Spectre and Virtuoso are registered trademarks of Cadence Design Systems, Inc). The transmission lines and passive baluns used in the 63.8 GHz-IC were simulated in the Sonnet Professional 2.5D field simulator.

5.1. The benchmark circuit: VCO with integrated switch

To evaluate the efficiency of the SILO approach, a conventional circuit using a VCO with wide tuning range and an output switch was designed. The system with the manufactured IC is shown in Fig. 9.

The schematic of the VCO can be seen in Fig. 10, together with the half-circuit of the designed output switch.

Parameter	$C^*_{var,min}$	$C^*_{var,max}$	$C_{var,min}$	$C_{var,max}$	L_B	R_E	R_{CC}
Value	65 fF	200 fF	145 fF	455 fF	0.41 nH	200 Ω	200 Ω
Parameter	C_1	C_2	C_3	C_4	V_{Bias}	V_{CC}	V_{tune}
Value	700 fF	200 fF	300 fF	300 fF	1.8 V	3.3 V	0 to 4 V

Table 1. VCO component parameters

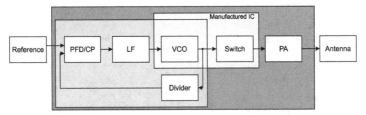

Figure 9. Pulsed frequency modulated continuous wave synthesizer system concept using an output switch

The VCO is based on a common collector Colpitts oscillator design, including a second varactor diode pair at the transistor base. It is described in detail in [8]. A short overview is given in the following.

A bipolar current mirror is used to drive the oscillator core. The emitter follower output buffer from [8] was replaced by a differential pair to increase common-mode rejection. The VCO frequency defining series resonant circuit consists of L_B and C_{in}:

$$f_{res} = \frac{1}{2\pi\sqrt{L_b C_{in}}} \tag{16}$$

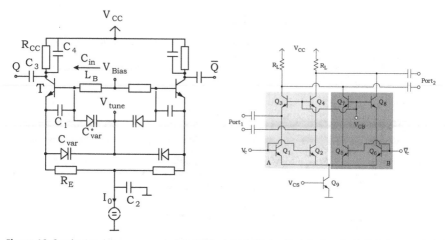

Figure 10. Synthesizer key components; left: VCO, right: half-circuit of single-pole double-throw switch

L_B is realized as a spiral inductor without tuning capability. Tuning is available by varying C_{in}, which has to be tuned over a wide tuning range using variable MOS-capacitance circuits.

For a minimum influence on the tuning range, C_P has to be minimized. It consists mainly of the collector base capacitance C_{CB} of transistor T and thus is given by size and bias conditions. C_S, which is determined mainly by C_{BE}, has to be maximized. Additionally, both varactor capacitance ranges have to be maximized. For a more detailed discussion, refer to [15]

The proposed pulsed ultra-wideband signal generation requires a switch after the frequency synthesizing PLL. The switch should have a minimum switching time in both on and off direction to enable the usage of very short pulses (in the $1 - 10\,\text{ns}$ range). Additionally, a constant input port impedance is important in order not to change the loading of the oscillator.

A switch circuit was designed based on [10]. The original work was aimed at a $22 - 29\,\text{GHz}$ UWB radar for automotive applications. Fig. 10, right, shows the half-circuit.

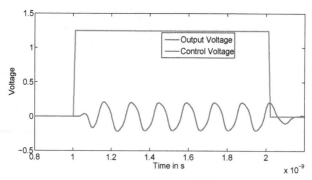

Figure 11. Switch transient simulation: Output voltage signal (blue) in reaction to control voltage (red) change.

The circuit works by switching the bias currents through branches A and B, implemented by transistors Q_1 to Q_4 and Q_5 to Q_8, respectively. This is done by alternating the control voltages applied to switching stages Q_1/Q_2 and Q_5/Q_6. The differential common base stages (Q_3/Q_4 and Q_7/Q_8) provide amplification and isolation, depending on the bias current. Transistor Q_9 provides the bias current, which is switched between the branches.

Fig. 11 shows the transient simulation of the output signal for a single rising V_C edge with a rise time of 5 ps. The delay between the control edge and a 90% of the output is below 250 ps. The addition of a matching network would improve insertion loss, but at the cost of worse area efficiency. The simulated input-referred noise was between $2.83 \, \text{nV} / \sqrt{\text{Hz}}$ and $3.67 \, \text{nV} / \sqrt{\text{Hz}}$.

A combination of VCO and output switch was simulated and then manufactured.

5.2. SILO oscillator concepts

As the injection locking property is universally stemming from oscillator theory, any oscillator can in theory be employed for switched-injection locking. There is an interesting trade-off to be made when considering an oscillator configuration for SILO building: The oscillator Q-factor should be high and excess loop gain should be low for better phase noise performance on the one hand, but a high-Q oscillator with low excess loop gain takes longer to begin oscillation, which is critical for pulsed angle modulated signal generation. A careful balance between the two qualities has to be found.

Another consideration has to be put into the point in the oscillator loop where the signal is injected into. In a cross-coupled oscillator, the resonator and gain stages are directly connected to the output. This means that there has to be a buffering circuit for the injected signal which provides backward isolation, in order to ensure the oscillation frequency of the oscillator is not influenced by the circuitry connected to the tank.

For the design of the SILO circuits, we concentrated on resonator-based oscillators, as they typically show better phase noise performance than inverter-based ring oscillators. A demonstrator implementation in discrete components was used for initial experimentation and verification of the viability of our approach. This circuit was aimed at a frequency range of 6 to 8 GHz. Subsequently, a SILO IC based on a pulse generator and a cross-coupled LC-oscillator was designed and manufactured. In a final step, a harmonics generator was combined with a Colpitts oscillator to sample a 5.8 GHz-signal and emit a 63.8 GHz-signal.

5.3. 6 and 8 GHz SMT SILO

For reference and for first experiments, SILO implementations based on surface mounted planar technology were realized. They are based on an ordinary common-collector Colpitts oscillator and designed for a natural frequency of 6 GHz respectively 7.5 GHz. In order to implement injection-locking, a directional coupler was added to apply the injection signal to the oscillator's output (see Fig. 12). The maximum achievable (10 dB) bandwidth is about 600 MHz at 7.5 ns pulse width.

Apart from parasitic technological limitations of lumped planar implementations, the single-ended design features an inherent source of self-locking to a harmonic of the switched power supply. Therefore, the pulse width is limited to about 10 to 20 ns in order to achieve a good compromise between bandwidth and minimum injection level. In consequence,

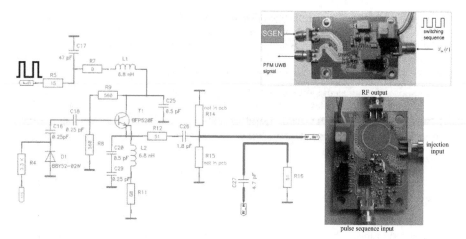

Figure 12. SILO SMT implementation; left: schematic of 7.5 GHz version; upper right: 7.5 GHz implementation; lower right: 6 GHz implementation

differential integrated circuit implementations are expected to deliver a significantly better self-locking suppression allowing much shorter pulsed in the order of 1 ns with comparable performance.

5.4. 7 GHz integrated circuit

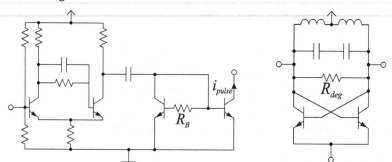

Figure 13. 7 GHz SILO circuits; left: Schmitt-trigger with peak generator, right: VCO with Q-degeneration resistor

The circuit consists of two active baluns for single-ended to differential and differential to single-ended conversion, a Schmitt-trigger with modified current mirror load for current peak generation and a simple cross-coupled oscillator for signal generation. It has an externally controllable pulse repetition rate and a pulse duration of approx. 1 ns. During operation it consumes 33 mA at 3.3 V supply voltage, while generating a > 330 mV$_{pp}$ signal. The generated signal has a 10 dB-bandwidth of over 2 GHz at 7.5 GHz center frequency.

Both Schmitt trigger with current peak generator and VCO with Q-degeneration circuits are shown in Fig. 13.

As efficient integrated circuits are built in a differential configuration but external circuitry and measurement equipment usually are only available in single-ended configuration, single-ended to differential (S2D) and differential to single-ended (D2S) conversion circuits are needed in the IC. We designed a simple active balun circuit that can act as both S2D-and D2S-converter. When employed as a S2D-converter, both outputs and one input are connected, when used as a D2S-converter, one output and both inputs are connected.

In order to control the pulse repetition rate externally, a Schmitt-trigger circuit with current peak generator was designed based on [13]. The circuit enables a wide variety of pulse repetition rates (1 − 80 MHz could be achieved with the measurement equipment at hand). The resistor R_B together with base-emitter capacitance C_{BE3} controls the time constant $\tau_{current}$ of the charging circuit:

$$\tau_{current} = R_B C_{BE3}. \tag{17}$$

The peak generator was designed for a pulse duration of 1 ns by selecting the size of the resistor $R_B = 5\,\text{k}\Omega$.

For the oscillator, a simple cross-coupled topology was chosen. As the oscillator has to lock to the injected phase, a low Q is preferable. In order to degenerate the Q, a resistor was connected in parallel to the LC-tank circuit. The current is provided by the peak generator. Fig. 13 shows the implementation.

A simple common-collector circuit is used as an output buffer to drive the 50 Ω load.

5.5. 63 GHz integrated circuit

The system developed for pulsed angle modulated signal generation at mm-wave frequency is shown in Fig. 14. The input signal of 5.8 GHz is coupled into the harmonics generator, which consists of a bipolar transistor with a resonant load. The load consisting of a transmission line of inductance L_1 and capacitors C_1 and C_2 is designed to couple the wanted 11th harmonic into the transformer. Fig. 17 shows the output power for the 1st, 10th, 11th and 12th harmonic depending on the input power. For an input power $> -3\,\text{dBm}$, the 11th harmonic is the strongest. The now differential signal is used to lock the VCO shown in Fig. 15.

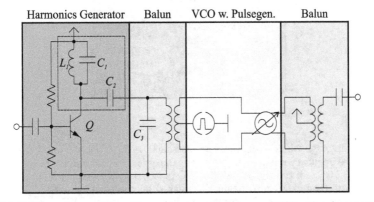

Figure 14. 63 GHz-system consisting of harmonics generator, baluns and VCO with pulse generator

The signal is coupled to the collector load transmission lines of the Colpitts oscillator using a transformer with a center tap. The center tap is connected to the pulsed current source of the oscillator.

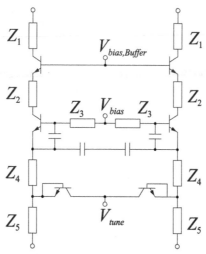

Figure 15. 63.8 GHz Colpitts voltage controlled oscillator schematic. Z_1 to Z_5 denote transmission lines

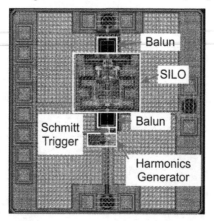

Figure 16. Layout of 63.8 GHz SILO

The simulation of the whole system was not possible. This is due to the fact that the system works in three frequency ranges, which differ by the order of magnitudes: The 5.8 GHz input signal, the 63.8 GHz output signal and the SILO pulse repetition frequency (10 − 100 MHz). Combined with the unknown modeling of switched injection-locking in the EDA software made it more viable to design each component (harmonics generator, VCO, pulse generator) separately. 16 shows the layout of the SILO circuit with its sub-components.

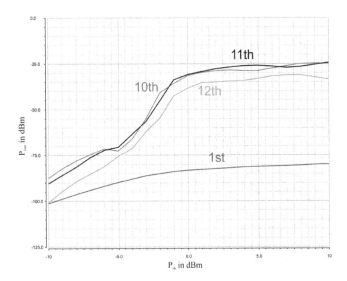

Figure 17. Power of the generated harmonics over the input power of the harmonics generator

6. Measurement setup and results

6.1. Verification of sampling theory

In order to verify the theoretical predictions concerning the switched injection locked harmonic sampling approach according to section 3.3, a demonstrator based on lumped planar components was built (see Fig. 18 and 19). It consists of a 480 MHz, 0 dBm signal source, a 10 MHz DAC modulated phase shifter, a single biased bipolar transistor frequency multiplier, a band pass filter (200 MHz @ 5.8 GHz) and the 5.8 GHz switched injection locked oscillator, which is turned on and off by the digital baseband synchronously to DAC modulation. Fig. 20 depicts the spectrum at the SILO's output. It features the typical sinc shaped peak comb in pulsed mode, which is aligned to and follows the injection frequency of 5.76 GHz when changed. When tuning the oscillators natural frequency (which is according to Fig. (20) different from the injection frequency) using a varactor diode, the sinc shape of the spectrum moves on the frequency axis while the peak positions do not change. These results prove most of the main claims of the generalized sampling theory according to (5) [4].

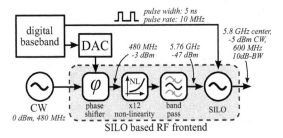

Figure 18. Implementation of communication and radar signal generator [4]

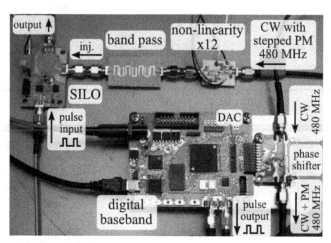

Figure 19. Hardware components for the 6 GHz transmitter system demonstrator (using lumped planar components SILO implementation)

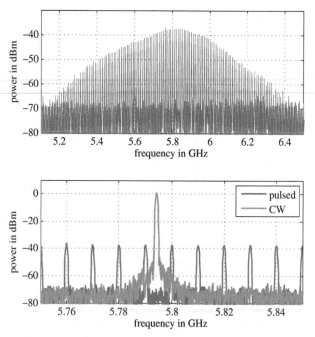

Figure 20. Spectrum of SILO based demonstrator with CW modulation; large peak: oscillator permanently on, comb: pulsed oscillator, background: comb zoomed out to show envelope, span 1.5 GHz [4]

6.2. Synthesis of communication signals

The synthesis of time domain communication signals was demonstrated using an 8 PSK modulation with cyclic transmission of all symbol values and maximum symbol rate, i.e. one symbol per pulse. The output signal of the demonstrator (Fig. 18, 19) was mixed to baseband using a quadrature mixer and displayed using an oscilloscope. Its waveform (Fig. 21) clearly shows the phase states and their repeatability in the IQ diagram. These results prove for the first time that it is feasible to generate UWB signals with more complex phase modulation than BPSK while at the same time keeping complexity and power consumption low.

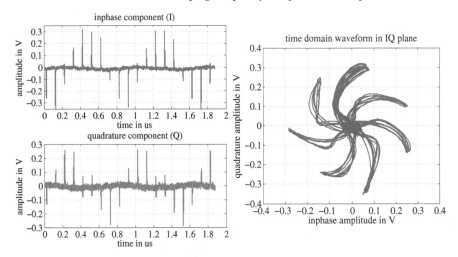

Figure 21. Generator output with 8 PSK modulation mixed to DC; left: inphase and quadrature component, right: IQ diagram [4]

6.3. Synthesis of radar signals

According to sections 3.5 and 4.3, the same simple hardware implementation used for communication signal synthesis (Fig. (18), (19)) can be employed to generate pulsed frequency modulated radar signals by repeatedly transmitting a limited list of phase samples. For a pulse rate of 10 MHz and a ramp slope of $20\,\text{MHz}/\mu s$, only 50 phase samples (one per pulse) are sufficient.

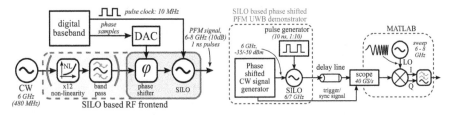

Figure 22. Measurement setup (on-waver) for the synthesis of radar signals using an integrated circuit SILO [3]

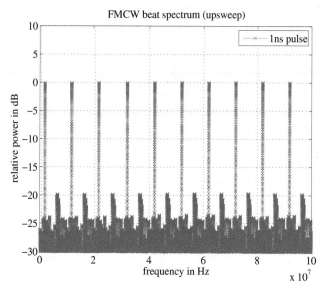

Figure 23. Beat spectrum (6-8 GHz SILO chip) of measured radar signal after mixing with linear sweep and before low pass filtering [3]

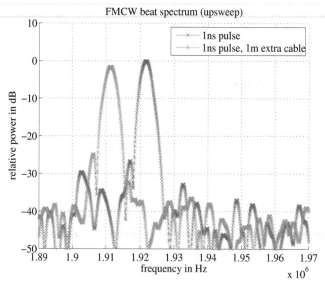

Figure 24. Zoomed beat spectrum (6-8 GHz SILO chip), comparison of two waveforms with different transmission delays [3]

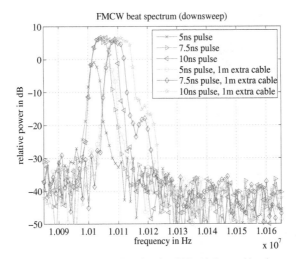

Figure 25. Comparison of 6-8 GHz chip (2 GHz bandwidth) with lumped implementation (wider pulses, smaller bandwidth) [3]

For verification, this approach was realized both using the previously employed lumped components SILO (6 GHz, 600 MHz bandwidth) and the first large bandwidth integrated circuit implementations (7 GHz, >2 GHz bandwidth) in order to demonstrate the resolution benefit for ranging. The setup for both experiments is depicted in Fig. 22; the generated and delayed signal is acquired using an oscilloscope and evaluated on a PC using a numerical computation software where it is mixed with a linear FMCW signal and analyzed in frequency domain (FFT).

Fig. 23 and 24 show the resulting beat frequency spectrum for the integrated circuit implementation using 1 ns pulses and 10 MHz pulse repetition rate. It corresponds to equation (10) except the small peaks that result from imperfections in the oscillator design leading to a slight turn-on pulse self-locking effect. Future designs are expected to fix this issue.

Comparing the results of the lumped and integrated circuit implementations (see Fig. 25), the benefit of much higher bandwidths regarding resolution becomes obvious. If the oscillator's spectral bandwidth is too small in relation to the sweep bandwidth, the beat frequency peak is broadened because of additional windowing through the narrowband SILO spectrum. Therefore, the oscillator bandwidth / pulse width should be adjusted to the desired sweep bandwidth in order to maximize spectral efficiency [3].

6.4. VCO with switch IC

The manufactured circuit is depicted in Fig.26. It measures $710 \times 1455\ \mu m^2$. For reasons of nonavailability of differential equipment, all measurements were done single-ended with the unused output terminated to ground with a 50 Ω resistor.

Fig. 27 shows the output power over the tuning range. The 10 dB decrease of output power compared to the previously published [8] VCO is attributed to the different VCO output buffer and the insertion loss of the switch.

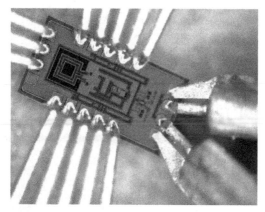

Figure 26. VCO with switch circuit IC photograph

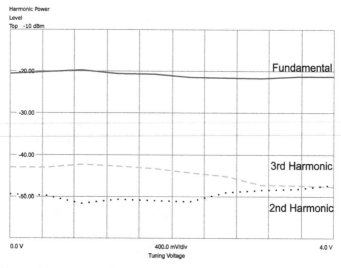

Figure 27. Measured output power of 1st, 2nd and 3rd harmonic of VCO with switch

The phase noise performance of the VCO with switch has deteriorated significantly from the previous [8] stand-alone VCO. This is mainly attributed to the new buffer structure which performed worse than anticipated.

6.5. 7 GHz SILO IC

The IHP Technologies SGB25V 250 nm SiGe:C BiCMOS process was chosen for manufacturing. It provides a cheap and flexible platform including one or two thick top metal layers consisting of aluminum. The advantage of using a BiCMOS process for a transmitter circuit is the possibility to build a system-on-a-chip (SoC) solution that integrates

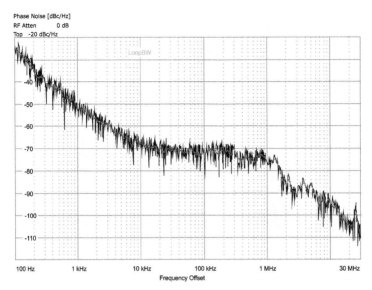

Figure 28. Measured phase noise of VCO with switch circuit

Figure 29. Manufactured 7 GHz SILO IC

digital baseband and analog RF circuits. Fig. 29 shows a chip photograph with connected measurement probes.

Fig. 30 shows the output power spectrum of the manufactured SILO.

The 10 dB-bandwidth stretches from 5 to 8 GHz. A single pulse is shown in Fig. 31. A single cycle of oscillator start up, oscillation and decay has a duration of 1.5 ns.

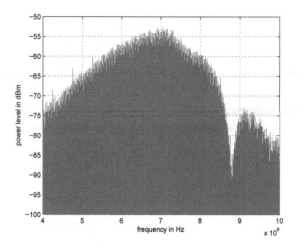

Figure 30. Measured output spectrum of 7 GHz SILO IC

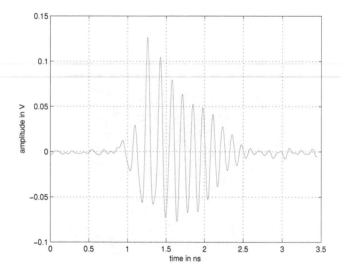

Figure 31. Transient output of 7 GHz SILO IC

7. Future work

Since this project is still ongoing, future work will cover further aspects that enhance theory and hardware implementation. Regarding pulsed angle modulated signals, more complex modulation schemes will be developed in conjunction with a more comprehensive study of error sources and their compensation. Furthermore, the first designs of the SILO circuit will be refined for an even better performance and higher integration level. Last but not least, hardware concepts for receiver technology are being developed.

Acknowledgement

This work was supported by the German Research Foundation (DFG - priority program SPP1202, grant VO 1453/3-2) within the project "Components and concepts for low-power mm-wave pulsed angle modulated ultra wideband communication and ranging".

Author details

Alexander Esswein and Robert Weigel
Institute for Electronics Engineering, University of Erlangen-Nuremberg, Germany

Christian Carlowitz and Martin Vossiek
Institute of Microwaves and Photonics, University of Erlangen-Nuremberg, Germany

8. References

[1] Barrett, T. W. [2000]. History of UltraWideBand (UWB) Radar & Communications: Pioneers and Innovators.

[2] Carlowitz, C., Esswein, A., Weigel, R. & Vossiek, M. [2012a]. A low power Pulse Frequency Modulated UWB radar transmitter concept based on switched injection locked harmonic sampling, *Microwave Conference (GeMiC), 2012 The 7th German*, pp. 1 –4.

[3] Carlowitz, C., Esswein, A., Weigel, R. & Vossiek, M. [2012b]. Synthesis of Pulsed Frequency Modulated Ultra Wideband Radar Signals Based on Stepped Phase Shifting, *IEEE International Conference on Ultra-Wideband (ICUWB)* .

[4] Carlowitz, C. & Vossiek, M. [2012]. Synthesis of Angle Modulated Ultra Wideband Signals Based on Regenerative Sampling, *IEEE International Microwave Symposium 2012*.

[5] Chandrakasan, A., Lee, F., Wentzloff, D., Sze, V., Ginsburg, B., Mercier, P., Daly, D. & Blazquez, R. [2009]. Low-Power Impulse UWB Architectures and Circuits, *Proceedings of the IEEE* 97(2): 332–352.

[6] Deparis, N., Loyez, C., Rolland, N. & Rolland, P.-A. [2008]. UWB in Millimeter Wave Band With Pulsed ILO, *Circuits and Systems II: Express Briefs, IEEE Transactions on* 55(4): 339 –343.

[7] Deparis, N., Siligarisy, A., Vincent, P. & Rolland, N. [2009]. A 2 pJ/bit pulsed ILO UWB transmitter at 60 GHz in 65-nm CMOS-SOI, *Ultra-Wideband, 2009. ICUWB 2009. IEEE International Conference on*, pp. 113 –117.

[8] Esswein, A., Dehm-Andone, G., Weigel, R., Aleksieieva, A. & Vossiek, M. [2010]. A low phase-noise SiGe Colpitts VCO with wide tuning range for UWB applications, *Wireless Technology Conference (EuWIT), 2010 European*, pp. 229 –232.

[9] Grass, E., Siaud, I., Glisic, S., Ehrig, M., Sun, Y., Lehmann, J., Hamon, M., Ulmer-Moll, A., Pagani, P., Kraemer, R. & Scheytt, C. [2008]. Asymmetric dual-band UWB / 60 GHz demonstrator, *Personal, Indoor and Mobile Radio Communications, 2008. PIMRC 2008. IEEE 19th International Symposium on*, pp. 1 –6.

[10] Hancock, T. & Rebeiz, G. [2005]. Design and Analysis of a 70-ps SiGe Differential RF Switch, *Microwave Theory and Techniques, IEEE Transactions on* 53(7): 2403 – 2410.

[11] J. Ryckaert et al [2006]. A 16mA UWB 3-to-5GHz 20Mpulses/s Quadrature Analog Correlation Receiver in 0.18/spl mu/m CMOS, *Solid-State Circuits Conference, 2006. ISSCC 2006. Digest of Technical Papers. IEEE International*, pp. 368 –377.

[12] Kohno, R. [2008]. Latest regulation and R&D for UWB inter-vehicle radar in millimeter wave band, *8th International Conference on ITS Telecommunications, ITST* .

[13] Lin, D., Schleicher, B., Trasser, A. & Schumacher, H. [2011]. Si/SiGe HBT UWB impulse generator tunable to FCC, ECC and Japanese spectral masks, *Radio and Wireless Symposium (RWS), 2011 IEEE*, pp. 66 –69.

[14] Miesen, R., Ebelt, R., Kirsch, F., Schaefer, T., Li, G., Wang, H. & Vossiek, M. [2011]. Where is the Tag? History, Modern Concepts, and Applications of Locatable RFIDs, *IEEE Microwave Magazine* . to be published.

[15] Pohl, N., Rein, H.-M., Musch, T., Aufinger, K. & Hausner, J. [2009]. SiGe Bipolar VCO With Ultra-Wide Tuning Range at 80 GHz Center Frequency, 44(10): 2655–2662.

[16] Roehr, S., Gulden, P. & Vossiek, M. [2008]. Precise Distance and Velocity Measurement for Real Time Locating Using a Frequency Modulated Continuous Wave Secondary Radar Approach, *IEEE Transactions on Microwave Theory and Techniques* 56(10): 2329–2339.

[17] Sewiolo, B., Hartmann, M., Guenther, O. & Weigel, R. [2006]. System Simulation of a 79 GHz UWB-Pulse Radar Transceiver Front-End for Automotive Applications, *VDE / ITG Diskussionssitzung Antennen und Messverfahren fuer Ultra-Wide-Band(UWB)-Systeme (UWB 2006) Kamp-Lintfort, Germany* .

[18] Trotta, S., Dehlink, B., Knapp, H., Aufinger, K., Meister, T., Bock, J., Simburger, W. & Scholtz, A. [2007]. SiGe Circuits for Spread Spectrum Automotive Radar, *Ultra-Wideband, 2007. ICUWB 2007. IEEE International Conference on*, pp. 523 –528.

[19] Vossiek, M. & Gulden, P. [2008]. The Switched Injection-Locked Oscillator: A Novel Versatile Concept for Wireless Transponder and Localization Systems, *Microwave Theory and Techniques, IEEE Transactions on* 56(4): 859 –866.

[20] Wentzloff, D. & Chandrakasan, A. [2006]. Gaussian pulse Generators for subbanded ultra-wideband transmitters, *Microwave Theory and Techniques, IEEE Transactions on* 54(4): 1647 – 1655.

ultraMEDIS – Ultra-Wideband Sensing in Medicine

Ingrid Hilger, Katja Dahlke, Gabriella Rimkus, Christiane Geyer, Frank Seifert, Olaf Kosch, Florian Thiel, Matthias Hein, Francesco Scotto di Clemente, Ulrich Schwarz, Marko Helbig and Jürgen Sachs

Additional information is available at the end of the chapter

1. Introduction

The exploitation of electromagnetic interaction with matter specifically with organic tissues is a powerful method to extract information about the state of biological objects in a fast, continuous and non-destructive (i.e. painless) way. These interactions are mainly based on two groups of phenomena.

One proceeds on an atomic and molecular level, which is typically described by the macroscopic quantities permittivity ε, permeability μ and conductivity σ. The physical reasons of possible interactions may be quite manifold. Here, in connection with ultra-wideband sounding, we restrict ourselves to pure electric interactions which affect the permittivity and conductivity via the motion of free charge carriers (free electrons and ions), the Maxwell-Wagner polarization (also Maxwell-Wagner-Sillars polarization) at boundaries, reordering of dipolar molecules or oscillations on an atomic or nuclear level. We assume that all involved substances have the permeability of vacuum $\mu = \mu_0$. An overview of relevant interaction mechanisms for biological tissue is given in [1], and sub-chapter 3 deals with some selected examples. The related effects are scattered over a huge frequency band covering 15...18 decades. In this paper, we limit ourselves to RF and lower microwave frequencies. Water – the key building block of life –shows dipole relaxation within the considered frequency band. Additionally, it has a very high permittivity in comparison with other natural substances. Hence, water will play an important role for UWB-sounding of biological tissue or human and animal subjects. Examples exploiting this fact are discussed in sub-chapter 5 dealing with breast cancer detection or in [2], which refers to lung edema. The frequency bands of our experiments were selected depending on physical requirements (propagation attenuation, relaxation time) and implementation issues of the sensor electrodes (e.g. antennas).

The second group of phenomena refers to macroscopic effects like reflection and refraction of electromagnetic waves. These effects appear at boundaries between substances of different permittivity or conductivity. Thus, a human body illuminated by radio waves will generate new waves which may be registered by an UWB radar sensor. The strongest waves are provoked by the skin reflection due to the large contrast between air and skin. But also inner organs will leave a trace in the scattered waves since firstly, electromagnetic waves within the lower GHz range may penetrate the body, and secondly, the various organs have different permittivity (e.g. due to different water content) leading to reflections at the organ boundaries. These waves can be used to reconstruct high resolution 3D microwave images of external or internal body structures and also to track their shape variation and motion.

It should be emphasized that motion is a strong indicator of vital activities like breathing, heartbeat or walking which can be registered remotely via UWB-radar sensing. This opens up new approaches of medical supervision as exemplified in sub-chapter 4, rescue of people in dangerous situation [3], [4] or supervision of people in need [5], [6].

In what follows, we like to review first some important requirements and technical solutions of high-resolution short-range UWB-sensor aimed at medical applications before we discuss a couple of selected aspects of medical ultra-wideband sensing in greater detail as for example:

a. Impedance (or dielectric) spectroscopy: It is performed to quantify and qualify biological tissue and cells. Here, we have to deal with reflection measurements at an open ended coaxial probe which is in direct contact to the material under test.
b. Organ motion tracking: It is aimed to observe temporal shape variations of the heart and the lung in order to trigger a magnetic resonance (MRI)-tomography. This task requires a remotely operating MIMO-antenna array with an up-date rate which is sufficiently high to follow mechanical motions up to 200 Hz.
c. Remote microwave imaging for surface reconstruction: It may be used as first step in a chain of further UWB-investigations based on remote sensing for inner organ evaluation. In the scope of this work, the data capture was implemented by scanning a torso. Under real conditions, such measurements must be made in real time using a large MIMO array (large in the sense of the number of antennas) in order to avoid artifacts due to body motions during the scan time.
d. Contact-based microwave imaging: In this case, the antennas are placed onto the skin either directly or through a thin dielectric layer in order to emphasize the radiation into the body. Applications are cancer detection or organ supervision and monitoring requiring densely occupied MIMO-arrays based on small radiators.

2. ultraMedis sensing devices

2.1. Requirements

The following overview summarizes some technical key features and requirements to be satisfied by the sensor electronics corresponding to the application types a) – d).

Bandwidth: UWB sensing is an indirect measurement method. As a general rule of thumb, one can state that the quantity respectively the reliability of the gathered information increases with the bandwidth of the sounding signal. It is predetermined and limited by the physical effects involved as well as technical implementation issues. In the case of impedance spectroscopy (application type a)), we applied Network Analyzers or M-sequence devices (see below) whose operational band was spanned from several hundred KHz to some GHz. For UWB-radar experiments, the frequency band was typically limited to 1-13 GHz or to 1-4...8 GHz. The lower cut-off frequency is typically determined by the size of the antennas while wave penetration into the body limits the upper frequencies. Correspondingly, the first frequency band was applied for application type c) which involves only propagation in air. The sensor device was a modified M-Sequence radar [7], [8]. If the sounding field must penetrate the body (applications b) and d)), the upper frequency may be reduced since wave attenuation dominates the other effects. Some details concerning the sensor structure are summarized in the next sub-chapter.

Field exposition: The strength of field exposition appearing in connection with UWB-sensing is usually harmless for biological tissue. Nevertheless, one should distinguish between an average charge and a short impact. A certain average charge of the test objects is required in order to achieve a given quality (in terms of signal-to-noise ratio) of the captured signals. The strength of the maximum impact is related to the type of sounding signals applied by the sensor. Sine waves and M-sequences cause maximum impacts of about the same strength as their average exposition is. In contrast, pulse systems lead to high-peak impacts even if their average charge keeps the same value as for sine waves or M-sequences. Hence, some attention should be paid to the selection of the sensor principle if applicators in direct contact with tissue and short electrode distances are involved (applications of type a) and d)) since this may lead to high field strengths within the test objects causing non-linear effects or even local damages.

Time stability: Here, the term 'time stability' refers to a summary of several facets of sensor performance like precision of equidistant sampling (i.e. linearity of time axis), long-term stability (drift), and short-term stability (jitter). These aspects pertain to all applications. They strongly affect the quality of the captured signals and, hence, the achievable results of the signal processing. In detail, the following items are concerned:

- the quality of time-frequency conversions via FFT, which is an important tool for signal processing
- the quality and durability of sensor calibration (3- or 8-term calibration),
- the limits of super-resolution techniques and the quality of background removal,
- the capability to detect weak targets in the neighborhood of strong ones, and
- the micro-Doppler sensitivity with respect to weak target detection and slow motions tracking.

Some additional aspects of this topic are summarized in Chapter 14. A thorough discussion of related problems and their linkage to the sensor principles is given in [9].

Measurement rate, channel number, data handling: Except for impedance spectroscopy, the applications mentioned above require MIMO-sensor arrays which have to run at a reasonable update rate. On the one hand, this assumes cascadable sensors in order to build multi-channel systems, and on the other it poses some challenges for the data handling resulting from the large number of measurement channels and the high measurement rate. Chapter 14 (section 2.1) adverts to some measures which avoid redundant and inefficient data. Irrespective of these measures, the data throughput will be still quite high so that standard PCs and interfaces quickly reach their capacity limits.

Radiators: The radiators represent the interface between sensor electronics and test object for applications b) – d). They have to convert guided signals into free waves and vice versa. As they are linear and time-invariant devices, they may be operated with any type of signals. Certainly, their major features are the bandwidth and the beam width which should be as large as possible if they are applied for UWB imaging. However, these characteristics describe their performance only insufficiently particularly for UWB short-range applications. Ideal UWB antennas for our purposes should provide a short and angular independent impulse response (time shape and wave front), they should convert the incoming signal completely into a free wave, and the incident fields should be converted into voltage signals avoiding any re-radiation or scattering by the antenna. These conditions are contradictory and cannot be met by a physically realizable antenna.

A short impulse response is needed for high range and image resolution as well as the ability to recover weak targets closely behind surfaces. Otherwise, we risk the loss of the target since a slowly decaying surface reflex distorts the target response. If that signal is too abundant, even sophisticated background removal strategies will not be able to dig it out.

The angular independent impulse response is important for the imaging algorithm. For every image pixel or voxel, it has to coherently integrate signals which are captured from different positions. In order to ensure the coherence of this integration, the propagation time to the considered pixel (voxel) must exactly be known. This involves the knowledge of the propagation speed as well as the knowledge of the deviation from a spherical wave front created by the antennas. In order to achieve a simple and manageable imaging algorithms, the involved antennas should avoid such distortions, hence they should be (electrically) small [9].

However, this contradicts the physical conditions for an efficient conversion between guided signals and freely propagating waves (see Bode-Fano limit and Chu-Wheeler limit [10]). Additionally, efficient antennas backscatter (re-radiate) half of the incident power in the ideal case. For targets in close proximity of the antennas, this leads to multiple reflections which are hardly to remove by signal processing. As we saw for the impulse behavior, the inefficient antennas behave again best regarding their re-radiations (structural antenna reflections are omitted here for shortness). Hence, one has to find a reasonable compromise between efficiency and impulse as well as scattering performance. Antenna efficiency is an important issue in connection with noise suppression and high path losses. For imaging at very close distances, noise induced measurement errors are falling below the strength of clutter and systematic deviations. Here, efficiency should take a back seat in antenna design in favor of a clear impulse response and low self-reflections. The sensitivity of the sensor electronics should compensate for the efficiency degradation of the antennas.

Furthermore, radiator related items concern array aspects such as the geometric shape of the array, radiator density (depending on antenna size and acceptable cross talk) and distribution within the array as well as polarimetric issues.

In the context of this chapter, we distinguish two types of antenna modes. For the first one, the antenna radiates in air, whereas the other mode refers to interfacial antennas which are in contact with the test object. In both cases, due to the short target distance, we have to deal with spherical waves and their specific reflection and refraction behavior which are accompanied by wave front deformations as well as the generation of evanescent and head waves [9].

Device miniaturization: The application of unusual radiators and the operation of dense MIMO-arrays require new sensor concepts avoiding long RF-cables (which have to be matched at both sides) as well as large and heavy measurement devices as network analyzers. Future MIMO-array implementations for medical microwave imaging should have jointly integrated radiator and sensor electronics in order to permit the operation of mismatched antennas, to increase the stability of the system, to reduce its susceptibility to environmental conditions (e.g. temperature variation or strong magnetic fields) and to simplify its handling. The project HaLos (Chapter 14) was addressing related questions of sensor integration.

2.2. ultraMEDIS sensor electronics

In view of the previous discussion, we mostly abstain from the use of network analyzers since they will not meet the requirements of future developments of the sensing technology even if they best fulfill the demands with respect to sensitivity, bandwidth and reliability of measurement data. A new sensor concept with comparable performance but higher measurement speed, better MIMO capability and integration friendly device layout exploits ultra-wideband pseudo-noise sequences (namely M-sequences) for the target stimulation instead of the sine waves of a network analyzer. This measurement approach was favored for our investigations. Device concepts applying sub-nanosecond pulses were rejected due to their inherent weakness concerning noise and jitter robustness. The interested reader is referred to Chapter 14 and [9] for further discussions of the pros and cons of various sensor principles.

The block schematics of the M-sequence prototype devices applied by ultraMEDIS are depicted in Figs. 4 and 6 in Chapter 14. The integrated RF key components were provided by the project HaloS while the implementation of prototype devices was performed by MEODAT GmbH and later on by ILMSENS. A special issue of an M-sequence device provides 12 GHz bandwidth. Its implementation is based on [8].

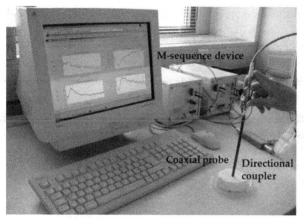

Figure 1. M-sequence based impedance spectroscopy (bandwidth 17 MHz – 4 GHz; 9th order M-sequence). Left: Device implementation with external coupler. Right: M-sequence device with internal coupler and rigid probe connection to improve measurement reliability.

Figure 4 of Chapter 14 (HaLoS-project) relates to the basic structure which can be found in all device modifications. Such device configurations were applied in an early project state for microwave imaging and organ motion tracking experiments. Involving a directional coupler, it is further used for impedance spectroscopy as exemplified in Fig. 1. Multi-channel systems and MIMO-arrays are based on the device philosophy as illustrated in Fig. 6 of Chapter 14. Implemented examples are depicted in Fig. 2 to Fig. 4.

Figure 2. M-sequence two-port network analyzer (operational band 40 MHz – 8 GHz, 9th order M-sequence, USB2 interface). It can be extended by an RF-switch matrix for MIMO-radar imaging.

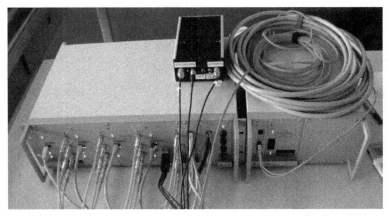

Figure 3. 4Tx-8 Rx MIMO device for organ motion tracking in MRI tomographs. (operational band 17 MHz – 4.5 GHz; 9th order M-sequence; maximum up data rate 530.4 Hz, Ethernet data link, data acquisition on Linux PC)

Figure 4. 8 Tx-16 Rx MIMO radar for microwave breast imaging (operational band 20 MHz – 6 GHz; 9th order M-sequence, USB2 interface). M-sequence units (as shown in Fig. 6 of Chapter 14) and RF-front ends are separated to get more flexibility for experimental purposes.

2.3. Antennas and sensor elements

2.3.1. Introduction

The exploitation of UWB microwave technologies for biomedical diagnostics requires the development of antennas and sensors tailored to this application. The integration of the antennas as a part of a complex system leads to serious compatibility and functionality constraints, which must be properly addressed for high system performance. Within ultraMEDIS, two goals were pursued: Firstly, the detection of early stage breast cancer and secondly, the development of a magnetic resonance imaging (MRI) compatible navigator system (Section 4). These two goals provide different challenges in terms of antenna design, implementation, and experimental evaluation, both with respect to mechanical and electrical constraints [10]. As both applications involve different approaches, they will be treated separately.

2.3.2. Dielectrically scaled antennas

For the process of detecting early stage breast cancer by means of microwave imaging (Section 5), the antenna size, the effective radiation of electromagnetic energy into the body, and the operational bandwidth turn out to be the main constraints regarding the design of the antenna.

The miniaturization of the antenna is of main concern to meet the requirements of the devised imaging technique (Section 5) of placing an array of many antennas surrounding the target under investigation (i.e. the human breast), considering also the small anatomic dimensions on the scale of the wavelengths of operation. In general, electrically small antennas are mismatched or narrowband [10], [11]. One possibility to overcome these obstacles is to use the antenna in *contact mode*, i.e., placing the antennas in contact with the target under investigation (e.g., the human body). With this *modus operandi* the antenna will radiate into a dielectric material (the human body), and it can be geometrically scaled by a factor of about $\sqrt{\varepsilon}$, where ε represents the dielectric permittivity of the target, without changing its electrical dimensions and, therefore, its radiative properties [12].

The *contact mode* presents advantages also with respect to the constraint of the effective radiation of electromagnetic energy into the body. In fact, it will not suffer from reflections occurring at the air-skin interface, due to the dielectric mismatch between the two grossly different media. This will also simplify the imaging processing since it prevents the need of surface reconstruction [10], [12]. Though, for practical and hygienic reasons, it is less convenient to put the array of antennas in direct contact with the patient's skin. However, the addition of a further layer, e.g. a disposable thin dielectric film, could spoil the effective radiation into the body[1]. Electromagnetic simulations (confirmed by measurements, Section 5.4) have shown that even the addition of a thin layer (~ 0.5 mm) can reduce the radiated power to less than half compared to the direct contact case. The implementation of a matching layer (with similar permittivity to human body tissues) can cure this effect (Fig. 5).

Eventually, particular attention has to be paid to the operational bandwidth of the antenna, especially to the lower cut-off frequency, which limits the penetration depth into the target. Based on a specific 14-layer model mimicking a trans-thoratic slice from the visual human data set, we have computed the penetration of electromagnetic waves into a human body, as shown in Fig. 6 [13], [14] and [15]. A strong increase of the signal attenuation with increasing frequency is clearly seen. Therefore, the lower cut-off frequency has to be set between 1 GHz and 2 GHz.

---

[1] The relevance of this phenomenon depends also on the antenna type used

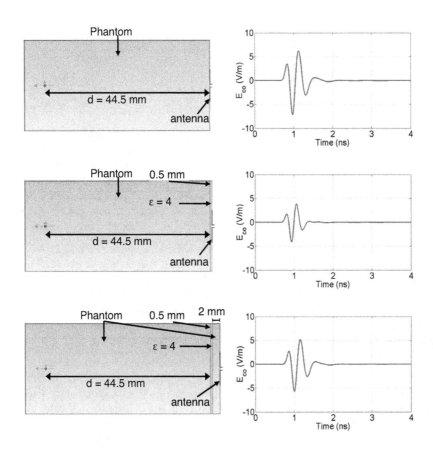

Figure 5. Simulated scenarios to investigate the effective radiation of electromagnetic energy into the body (on the left). The antenna used is a bow-tie excited by a Gaussian pulse of a duration of around 80 ps FWHM. The *Phantom* material is a homogenous dispersive material simulating the dielectric behavior of the human body tissues. The results (on the right) represent the time-dependent co-polar component of the electric field evaluated at a distance of 44.5 mm from the phantom interface (the green spot in the figure). The examined cases are, from top to bottom: the antenna in direct contact with the phantom; with the implementation of a thin dielectric layer; with the implementation of a matching layer plus a thin dielectric layer, respectively.

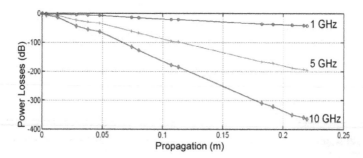

Figure 6. Penetration depth into a 14-layer model computed for different frequencies (see legend) [13].

The dielectric scaling of traveling-wave antennas, like tapered slot antennas and horn antennas, has to consider all three factors - the phase constant, the wave impedance, and the feeding structure [16]. Accordingly, extended iterative full-wave simulations were performed. The key parameters of the wideband radiation properties of double-ridged horn (DRH) antennas turned out to be the curvature of the ridges, the spacing between them, and the geometry of the pyramidal casing itself. Traveling-wave antennas have the benefit that the dielectric medium used to scale the antenna will serve as an embedded *matching medium*.

Our initial design was based on the immersion of dielectric liquids into the sinkhole of a dielectrically scaled DRH antenna [15]. The antenna could successfully be manufactured, using acetone as dielectric medium, with a scaling factor of about 4, but it was still insufficient to obtain a sufficiently compact antenna. The straight-forward approach to solve this problem was to replace the acetone by alternative high-permittivity dielectric materials, like a solid sintered ceramic. The ceramic powder is the commercial product LF-085 manufactured by MRA Laboratories based on neodymium titanate [17].

As the complex permittivity ε of the ceramic body of the antenna plays an important role not only in terms of antenna design but also in terms of *matching medium*, we performed measurements to access the complex permittivity following two different strategies: employing a split-post dielectric resonator (SPDR, [18]) and a dielectric resonator (DR) technique [19][20]. Both techniques are resonant methods and, hence, limit the experimental studies to a small set of discrete frequencies, because specific sample geometries are required for each measured frequency. The results showed that the sintered ceramic presents low frequency dispersion with a mean value of the real part of the permittivity $\varepsilon' \cong 72$ [21], offering the potential for a scale factor of around 8.

The exploitation of the full potential of dielectric scaling leads to an aperture size of only 11 mm × 16 mm, but also to the reduction of the input impedance by the same scaling factor as by which the geometrical dimensions are scaled, resulting in a low value below 10 Ω. This value implies a large mismatch in terms of standard electronic equipment, which is usually designed for a characteristic impedance of 50 Ω. In order to maintain the compatibility with standard electronic equipment, the antenna retains an aperture size of 24 mm × 24 mm, and a frequency bandwidth ranging from 1.5 to 5.5 GHz (Fig. 7).

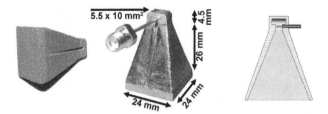

Figure 7. The left-hand panel shows the completely processed ceramic body of the antenna. The center panel depicts the final antenna including metallization, feed line and dimensions. The right-hand panel displays the profile of the ridges.

In order to properly manufacture the antenna and obtain good yield (of around 75 %) and reproducibility, specific manufacturing steps have to be devised, as indicated in the right image of Fig. 8 [22]. First, the white-colored ceramic powder (Fig. 8a) is pressed into the specific pyramidal shape; a cubic base accommodates the asymmetric feed (Fig. 8b). The dimensions of the raw body are slightly enlarged in order to respect the shrinkage upon sintering. The sintered body is complemented by grooves, which form the ridges (Fig. 8c), and is subsequently galvanically metalized with copper or gold (Fig. 8d). Due to the high permittivity of the ceramic body and a feed impedance of 50 Ω, the ridges are nearly linear in geometry, in contrast to the markedly curved shapes found in antennas for operation in air [12][15]. The feed is provided by a coaxial cable whose center conductor is fed through a small bore to the narrow end (diameter about 1.2 mm) of the ridged waveguide. A plastic housing and epoxy fixture provide a compact and mechanically rigid construction, to protect the ceramic body and the coaxial feed against torque and damages due to improper handling. It also provides a mechanical fixture to mount the antennas in an array of complex geometry (Fig. 8e). Further details of the manufacturing processes are given in [23].

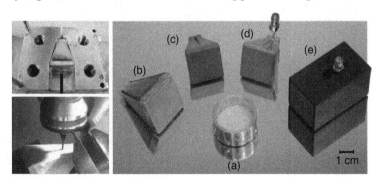

Figure 8. The left-top picture shows a dry pressed green body of the antenna with some lubricant on it inside the dismantled mold. The left-bottom picture depicts the milling process to work in the ridges into the sintered ceramic body. The right picture illustrates the manufacturing steps for the sintered horn antenna: powder raw material (a), pressed raw (b), sintered (c), metallized (d), and fully packaged versions (e).

A further issue of dielectrically scaled antennas is related to their measurement and characterization. As common measurement techniques and equipment cannot be applied, we followed three different strategies: measurements made in the frequency domain, measurements made in the time domain, and basic tests with volunteers.

In order to provide dielectrically matched surrounding conditions for the antenna body, the antennas were tested in de-ionized water. The results were then compared with data obtained in a more realistic environment, i.e. with the antenna put into contact with phantoms mimicking human tissues. The phantoms consisted of oil, water and some additives [24]. The dielectric permittivity ε' and the loss tangent $\varepsilon''/\varepsilon'$ of the phantoms can be controlled by changing the percentage of oil [24], [25] (Section 5.2).

The frequency behavior of the reflection coefficient is shown in Fig. 9. It can be observed that while the reflection coefficient for the test against the phantom (with 40% oil, Section 5.2) approaches levels around −8 dB, the antenna is even better matched in water, leading to a further decrease of the reflection coefficient by 4 dB in the frequency range of interest. The compromise between input matching to a certain medium and the geometrical dimensions of the antenna denotes the key trade-off exploited for our design. In order to study the reflection occurring at the aperture plane, which is influenced by the permittivity matching between the dielectric medium composing the antenna body and the human skin, we performed time domain reflectometry (TDR) measurements by having the antenna radiate into different media [22].

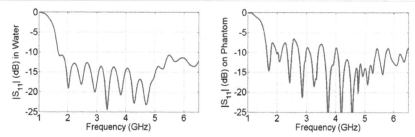

Figure 9. Reflection coefficient measured in water (left) and on a skin mimicking phantom (right).

The results for water and skin (in-vivo) are shown in Fig. 10. The amplitude of the reflected signal with the antenna operating in water is significantly smaller than the one with the antenna operating on skin. This result indicates, in agreement with the frequency domain measurements, that the antenna is better matched to water than to skin. We note from Fig. 10 that the reflection occurring at the aperture due to impedance mismatch results in a signal with a longer decaying time. The larger the impedance mismatch is the longer the decaying time is. This feature is due to the fact that part of the reflected energy does not leave the antenna through the well-matched feed towards the signal source but remains within the antenna body.

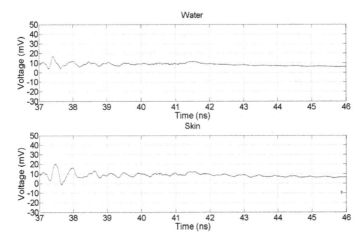

Figure 10. Time domain reflections caused by the aperture plane due to dielectric mismatch between the antenna and water and skin, respectively (top and bottom).

The measurement of the radiating behavior is more complicated. To evaluate standard antenna parameters (e.g. gain, radiation pattern, etc.), the antennas should be placed in the Fraunhofer region. However, due to the high dielectric losses of water, the antenna could be placed at a maximum distance of 10 cm, which is not sufficient to meet the Fraunhofer region (starting from around 35 cm), but still is large enough to let the antennas operate in the Fresnel region. Near-field measurements are of main concern since the antennas are designed to be used in contact mode for biomedical imaging applications. The results show that the antennas present a flat frequency response (measured along the boresight direction), after the compensation of the frequency dispersion of the water (left diagram in Fig. 11), and 3 dB beam widths of nearly 20° for the E-Plane, and nearly 28° for the H-Plane (right diagram in Fig. 11).

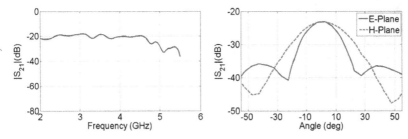

Figure 11. Transmission behavior between two identical antennas operating in de-ionized water at a distance of 10 cm. The frequency response, compensated for water frequency dispersion is shown in the left diagram, while the near-field pattern is represented on the right. The blue and red curves illustrate a cut along the E-Plane and the H-Plane, respectively.

Eventually, in order to demonstrate the functionality of the ceramic antennas under realistic conditions, we have performed preliminary transmission measurements through a breast of a female volunteer, and the monitoring of the heartbeat, as illustrated in Figs. 12 and 13. The dynamic range of the achievable signal can be determined from Fig. 12 in comparison with the ideal transmission through a 4 cm path inside distilled water by a face-to-face arrangement of the antennas. The monitoring of the heartbeat was performed on a 35 years old healthy male volunteer. During measurement, the volunteer was sitting still and was holding breath in the state of maximal breathing in. The measurement was performed with an M-sequence radar in a bi-static configuration [Section 2.2]. Figure 13 shows the heartbeat signals as monitored. Upon Fourier transformation, we extracted a beat rate of nearly 75 beats per minute, which is considered normal for an adult. The results display very clearly the characteristic feature of heartbeat, thus manifesting a favorable dynamic range. This opens up promising applications for realistic monitoring and imaging tasks. Further details of these tests can be found in [26] and [22].

The full dielectric scaling, as previously stated, offers the potential for a further size reduction of the antenna. Accordingly, we continued our research and succeeded in developing a ceramic DRH antenna with an aperture of only 16 mm × 11 mm and a lower cut-off frequency around 1.5 GHz. Due to the input impedance of the antenna below 10 Ω, active receive and transmit versions are under development in the framework of HALOS (Chapter 14), employing an UWB low-noise subtraction circuitry and power amplification [27]. The manufacture of the tiny antenna followed similar production steps as for the previous version. It proved quite challenging because of the reduced size, requiring additional specific production steps and iterative testing procedures [28].

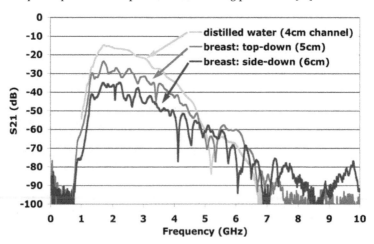

Figure 12. Transmission measurements through a female breast performed with the ceramic double-ridged horn antennas in comparison with a reference measurement of 4 cm distilled water (upper curve).

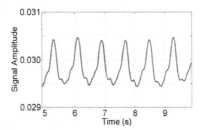

Figure 13. Monitored heartbeat signal of a 35 years old healthy male volunteer. Upon Fourier transformation, we extracted a beat rate of nearly 75 beats per minute.

2.3.3. MRI compatible antennas

Magnetic resonance imaging (MRI) systems are among the most sensitive diagnostic methods in medicine for the visualization of soft tissue [10]. At present, more than ten million MRI examinations of patients are performed per year worldwide. Given such a progressive development, further improvements of this diagnostic technique are under way. However, MRI systems are not *per se* capable of creating focused images of moving objects like the human heart or the thorax of the patient while breathing. Instead, additional techniques like breath holding, ECG triggering, or MR navigation methods are required. Such techniques either cause some inconvenience for the patient, or they are even not applicable for upcoming generations of MR scanners. A novel approach which overcomes these obstacles is the use of low-power multi-static UWB radar as a contactless navigator technology for MR tomography [29], [30]. To devise such navigators, the design of antennas compatible with MRI systems, i.e. antennas which do not interact neither electrically nor mechanically with the operation of the MR scan, is needed. The strategy to follow when designing an MRI compatible antenna is the minimization of mutual interaction between the metallized antennas and the strong static and gradient magnetic fields. Several experiments with conventional wideband antennas showed strong mechanical interactions, pointing out the need for special antenna designs [31]. The operational conditions inside an MR scanner are determined by three different types of fields. First, a static magnetic field B_{stat} = 1.5 to 7 T provides a reference orientation of the nuclear spins of the regions under inspection. Furthermore, gradient magnetic fields with a slope of dB_{grad}/dt = 50 T/s at the rising edge are switched during diagnostic measurements to allow for spatial (depth) information of the acquired molecular information. Finally, the nuclear spins are set into precession by a strong (KW range) RF signal at 42.58 MHz/T. The gradient fields induce eddy currents in the metallized sections of the antenna which, in turn, interact with the static magnetic field by exerting a mechanical torque on the antenna structure. The torque can reach peak values of the order of 0.045 Nm for a contiguous metallized area of 20 mm × 30 mm. This value is high enough to result in mechanical amplitudes of several millimeters, deforming or moving the antenna structure, especially in the case of mechanical resonances. Furthermore, the magnetic fields of the eddy currents can lead to artifacts of the MR-image. These interactions inhibit the beneficial application of UWB navigation for magnetic resonance imaging and, therefore, must be avoided. We used a 3-T MR scanner with the resulting RF frequency of 127.8 MHz, which is ten times smaller than the lower cut-off frequency of the UWB antennas employed. As the frequency response of a typical antenna corresponds to a high-pass filter of first order, the stop-band attenuation amounts to 20 dB per decade, indicating the risk of collecting RF power even in the presence of path-loss and shadowing.

The minimization of contiguous metallized area and, hence, eddy currents, is therefore the main issue to be addressed by the antenna design. Additional constraints arise from the intended applications in biomedical diagnostics: weakly frequency-dependent radiation patterns over the entire operational bandwidth, good decoupling between neighboring antennas, and a lower cut-off frequency around 1 GHz. The DRH antenna was identified to be a suitable UWB antenna type to accomplish these requirements. Due to the functional principle of DRH antennas, the minimization of contiguous metallized areas and the realization of a weakly frequency-dependent radiation pattern are in conflict with each other. Horn antennas are typically made entirely out of metallic parts of high electrical conductivity σ, thus suffering from the induction of eddy currents under MR-scanner conditions. Therefore, the major challenge was to modify the double-ridged horn antenna to achieve MR-compatibility, without compromising the favorable radiation properties.

Inspired by commercial counterparts of DRH antennas, we removed the H-plane sidewalls of the pyramidal horn, leaving just a thin wire in the plane of the aperture, as illustrated by the left picture in Fig. 14. As a result, the lower cut-off frequency could be reduced from 2.6 GHz to 1.5 GHz for otherwise unchanged dimensions and operation in air. The comparison with a conventional double-ridged horn antenna with a similar bandwidth revealed that this improvement was achieved at the expense of increased beam width, side-lobes and backward radiation, predominantly at frequencies below 3 GHz, due to the modified aperture distribution and diffraction at the edges of the open construction. The increased beam width led to a slightly increased crosstalk [32]. It can easily be compensated for by re-orienting the antennas relative to each other. While the crosstalk for conventional DRH antennas becomes small for an H-plane alignment, the MR-compatible versions have to be aligned along the E-plane due to the removed H-plane sidewalls and, thus, reduced shielding.

The thickness of the metallization was also reduced in order to exploit the skin effect for a decoupling of the low-frequency eddy current paths. The metal planes were replaced by metallized dielectric boards with a metallization thickness of 12 μm (Fig. 14). This value corresponds to about twice the skin depth at the lowest frequency used. The high-frequency currents determining the radiation of the antenna remain essentially undisturbed while the eddy currents in the KHz range are strongly attenuated by the high sheet resistance. For further optimization of the remaining metallized areas, the distribution of surface currents in the UWB frequency range was inspected by electromagnetic simulations (right image of Fig. 14). Typical results for the normalized surface current are illustrated at 5 GHz (left-most). The surface current is concentrated near the position of the ridge and the edges of the pyramidal frame. According to our expectations, the number of current loops was found to increase with frequency; in contrast, the current distribution across the backward cubical part of the antenna showed little frequency dependence. Based on these observations, a compromise was sought to reduce the plane metallization with the minimal possible distortion of the broadband current distribution. As a result, the conductor faces of the horn sections were separated into strip lines, straight and elliptically shaped, separated by 1 mm, and oriented parallel to the most common current paths, with plain connections at the face edges only. The central part of the right image of Fig. 14 illustrates the resulting geometric arrangement of the slots, while the normalized surface current of the modified antenna at 5 GHz is shown in the right-most part. The main features of the current distribution could be sustained qualitatively both on the pyramidal faces and the backward cubical part of the antenna. Differences occurred mainly for the currents oriented perpendicular to the slots. It is this minor change in current distribution which causes the modified radiation properties discussed above. The ridges themselves required special attention. A grid of holes was eventually identified as the proper solution to reduce the metallization area of the ridges without disturbing the high-frequency current distribution too much. In order to reduce the maximal loop size for low-frequency currents, the outer contour of the horn section was cut and shortened by standard surface-mounted device capacitors.

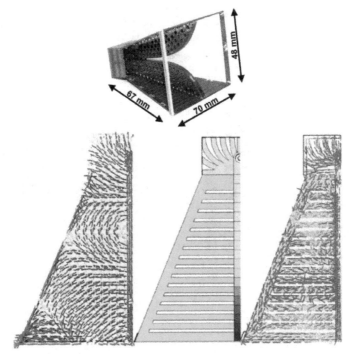

Figure 14. MR-compatible double-ridged horn antenna for a lower cut-off frequency of 1.5 GHz (top). The lower image shows the simulated, normalized current distribution of an unmodified DRH antenna at 5 GHz (left-hand part), the layout of the resulting MR-compatible DRH antenna (center part), and the current distribution of the modified DRH antenna at 5 GHz (right-hand part).

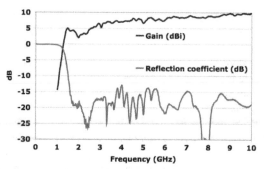

Figure 15. Measured reflection coefficient of the MR-compatible double-ridged horn antenna (lower curve) and the measured antenna gain (upper curve) versus frequency.

Fig. 15 displays measured results for the reflection coefficient and the gain of the modified DRH antenna. A return loss above 10 dB was achieved over the frequency range from 1.5 to 12 GHz. Radiation measurements in an anechoic chamber yielded the radiation patterns illustrated in Fig. 16 for two orthogonal cuts with respect to the plane of the ridges. The half-power beam width, indicated as the black contour line in Fig. 16, was found to vary between 30 and 50 degrees, thus covering a range suitable for the envisaged applications. Except for frequencies around 2 GHz, the main lobe showed little spectral variation. The corresponding frequency variation of the antenna gain is displayed in Fig. 16. These results were found in good agreement with the numerical simulations.

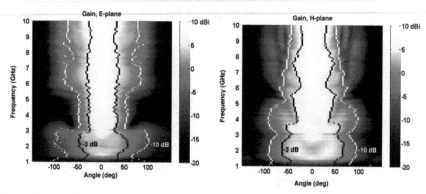

Figure 16. Two-dimensional representation of the measured radiation pattern of the MR-compatible double-ridged horn antenna for the E-plane (left) and the H-plane (right) through the main beam. The scales indicate the antenna gain in dBi. The black and white contour lines illustrate the corresponding beam widths at 3 and 10 dB below the frequency-dependent maximum gain, respectively.

The transient response of the antenna is shown in Fig. 17. Despite the open geometry of the MR-compatible antenna, a low signal distortion could be sustained. The slight angular dependence of the time responses can be attributed to an offset between the phase centers of the antennas and the center of rotation of the antenna positioning system.

The MRI compatible DRH antennas were implemented as part of a UWB MR navigator, by means of which it was possible to take images of the myocardium for the first time without using an ECG as navigator. The quality achieved was comparable with the one achievable with the ECG navigator (see Section 4).

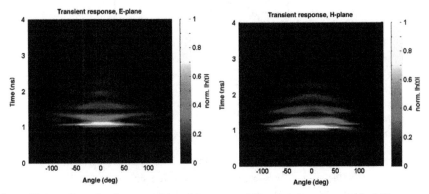

Figure 17. Two-dimensional representation of the measured time domain response of the MR-compatible DRH antenna for the E-plane (left) and the H-plane (right) through the main beam. The scales indicate the normalized impulse response of the antenna.

3. Weak electromagnetic fields and biological tissue

3.1. Impact on living cells

The electrical properties of biological tissues and cell suspensions have been of interest for over a century for many reasons. They determine the pathways of current flow through the body and are very important for the analysis of a wide range of biomedical applications such as functional electrical stimulation and the diagnosis and treatment of various physiological conditions with weak electric currents, radio-frequency hyperthermia, electrocardiography, and body composition. On a more fundamental level, the knowledge of these electrical properties can lead to an understanding of the underlying basic biological processes. Indeed, biological impedance studies have long been an important issue in electrophysiology and biophysics; interestingly, one of the first demonstrations of the existence of the cell membrane was based on dielectric studies on cell suspensions [33].

Biological tissues are a mixture of water, ions, and organic molecules organized in cells, sub-cellular structures, and membranes, and its dielectric properties are highly frequency-dependent in the range from Hz to GHz. The spectrum is characterized by three main dispersion regions referred to as α, β, and γ regions at low, intermediate, and high frequencies [34]. Biological materials can show large dispersions, especially at low frequencies (Fig. 18). Low frequencies are mainly caused by interfacial polarizations at the surfaces between the different materials of which a cell is composed [35]. Reviews of the dielectric properties of cells and the different dispersions are given in the literature [36], [37].

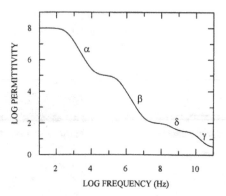

Figure 18. Spectrum of the dielectric properties of cell suspensions and tissues.

The step changes in ε_r are called dispersions and are due to the loss of particular polarization processes as frequency increases. The α-dispersion is due to the flow of ions across cell surfaces, the β-dispersion results from the charge at cell membranes, the δ-dispersion is produced by the rotation of macromolecular side-chains and bound water, and the γ-dispersion is due to the dipolar rotation of small molecules particularly water [35] (figure reproduced with permission from Elsevier).

When exposed to electric fields, living cells behave as tiny capacitors, accumulating charges on the cell surface. The permittivity of living cell suspensions is dependent on the frequency, and falls in a series of the dispersions described above, as frequency increases. The β-dispersion, between 0.1 and 100 MHz, results from the build-up of charges at cell membranes. The difference between permittivity measurements made at two frequencies, on either side of the β-dispersion range, is proportional to the viable biomass concentration. With spherical cells, the permittivity increment is given by equation [38].

$$\Delta\varepsilon = \frac{9\,P\,r\,C_m}{4} \tag{1}$$

As long as there is no change in the cell radius r or the membrane capacitance C_m, the permittivity increment $\Delta\varepsilon$ is proportional to the cell volume fraction P [39].

As a starting point for developing new applications, it is critical to characterize differences in the dielectric properties of the cells, for example human leukocyte subpopulations [40]. Even though, a comparative analysis of the dielectric properties of the cells is necessary, and the UWB radiation on cells itself has to be characterized, too. For this reason, experiments with two different cell lines (tumor cell line BT474 and fibroblasts BJ) were performed. Cell suspensions of these cell lines were disseminated, and the growth rate was determined. Afterwards, the cells were seeded on 96-well plates, cultivated for 24 h and exposed to UWB radiation *via* UWB-M-sequence radar with double-ridged horn antennas of about 10 dBi average gain for 5, 30 or 60 min. As non-treated control, for the same time, plates were placed in a Faraday cage (to avoid any irradiation). After continued incubation for 24, 48 and 72 h, the vitality of cells was determined by colorimetric identification (MTT assay for measuring the activity of enzymes that reduce MTT [3-(4.5-Dimethylthiazol-2-yl)-2.5-diphenyltetrazolium bromide, yellow tetrazole] to formazan, giving a purple color). The measured vitality of control cells was normalized to 100%, and the vitality of exposed cells was put into relation. The vitality of exposed cells was related to non-exposed cells. Due to biological fluctuations, data between 70% and 120% vitality were assessed as not influenced. As depicted in Fig. 19, none of the determined cells was influenced by ultra-wideband electromagnetic waves.

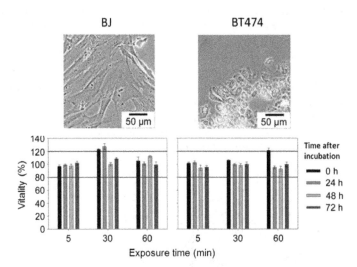

Figure 19. Impact of ultra-wideband electromagnetic waves on the vitality of living cells. The upper part of the figure shows light images of the fibroblast cell line BJ and the cancerous cell line BT474. The lower figure depicts the vitality of the fibroblasts BJ and the cancerous cells BT474 after UWB exposition with 4 mW for 5, 30 or 60 min. The vitality was observed 0, 24, 48 or 72 h after exposure. The depicted vitality of exposed cells is related to non-exposed cells. Due to biological fluctuations, data between 80% and 120% vitality was not considered to be cytotoxic [25].

3.2. Animal tissue

The electrical properties of tissues and cell suspensions are most unusual. They change with frequency in three distinct steps (dispersions as described above) and their dielectric constants reach enormous values at low frequencies. Extensive measurements were carried out over a broad frequency range extending from less than 1 Hz to many GHz. The observed frequency changes of these properties obey causality, i.e., the Kramers-Kronig relationships which relate changes of dielectric constants with conductivity changes. A number of mechanisms were identified which explain the observed data. These mechanisms reflect the various compartments of the biological material. These include membranes and their properties, biological macromolecules and fluid compartments inside and outside membranes [41]. Special topics include a summary of the significant advances in theories on counter ion polarization effects, dielectric properties of cancer *vs.* normal tissues, properties of low-water-content tissues [42], and macroscopic field-coupling considerations. The dielectric properties of tissues are often summarized as empirical correlations with tissue water contents in other compositional variables. The bulk electrical properties of tissues are needed for many bioengineering applications of electric fields or currents, and they provide insight into the basic mechanisms that govern the interaction of electric fields with tissue [43].

Using devices with our own configurations, the dielectric properties of different porcine and bovine tissues were determined [25]. Different measuring points were defined on the surface of udder, fat, liver, muscle, and kidney of porcine and bovine tissue (homogenous structure) and the permittivity of these points was measured three times (selected tissues in Fig. 20, left panel). Afterwards, the tissue under these measuring points was excised and dried to calculate the water content. Water content and permittivity ε' were related to each other, so we could clearly differentiate between fat, low-water-content tissue, with a low permittivity ($\varepsilon' \approx 8$) and liver, muscle or kidney ($\varepsilon' \approx 40$) as high-water-content tissues. The high-water-content tissues show similar permittivity ε' values whereas fat of porcine and bovine origin can be distinguished (Fig. 20, right panel).

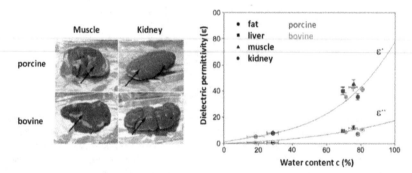

Figure 20. Determination of the dielectric properties of different porcine and bovine tissues at defined measuring points (left panel). Real and imaginary part of permittivity ε' at 2 GHz of porcine and bovine tissue in relation to the water content (right panel). The standard error represents six independent measurements [25].

3.3. Bacterial cell wall identification based on their dielectric properties

The identification of bacterial strains in biological media is a matter of interest in very different fields of modern life. Examples are in food hygiene and food industry, catering and gastronomy [44], [45], in environmental research activities, fermentation processes for the production of medical drugs, such as insulin, antibiotics, and other [46]-[48], and in the diagnosis of infections in clinical and veterinarian applications [49]. Depending on the respective research and application field, bacterial strains are currently detected by complex methods, for example: polymerase chain reaction (technique to amplify a single or a few copies of a piece of DNA), fluorescent *in situ* hybridization, DNA microarray and Raman-spectroscopy, etc.

Different studies have shed some light into the biomass determination of different microbial suspensions *via* dielectric spectroscopy. Mishima *et al.* investigated growth kinetics of bacterial, yeast and animal cells by dielectric monitoring in the frequency range of 10 kHz - 10 MHz [50]. The determination of bacterial growth by dielectric measurements was also shown by Harris *et al.* [51]. Jonsson *et al.* measured the concentration of bacterial cells *via* indirect methods based on the dielectric determination of ions in the suspension, which are released by killed cells [52]. Benoit *et al.* showed that it is possible to discriminate the hydrophobic or hydrophilic features of bacterial suspensions by determining the dielectric permittivity [53]. Nevertheless, no data are available for discrimination on the basis of bacterial structures *per se*, such as the presence of Gram-positive or Gram-negative bacterial strains in biological samples [54].

Therefore, two different Gram-positive bacterial strains (*Micrococcus luteus* and *Bacillus subtilis*) and two Gram-negative bacterial strains (*Escherichia coli* and *Serratia marcescens*) were cultivated under standard conditions using Standard I media and shaking flasks. Bacterial strains were incubated for 24 h at 37°C in an incubation shaker. To assess whether the Gram-status of bacteria could be determined by dielectric spectroscopy, bacterial suspensions were transferred to 50 ml tubes and centrifuged. The supernatant (liquid above precipitate) was removed, the pellet was washed in 0.9% sodium chloride solution and, finally, the dielectric properties of the bacterial biomass (pellet of 10 ml) were determined. Dielectric spectroscopy of bacterial strains and suspensions was performed using a network analyzer in a frequency range from 30 kHz to 6 GHz (HP 8753D) and a coaxial probe (High temperature probe). The real ε' and imaginary ε'' part of permittivity was determined in a frequency range from 50 MHz to 300 MHz [54].

In the frequency range between 50 and 300 MHz, dielectric spectroscopy revealed higher values of the real part of permittivity ($\varepsilon'_{(+)} \approx 160$ Gram-positive) of the Gram-positive bacterial strains *Micrococcus luteus* and *Bacillus subtilis* compared to the Gram-negative strains *Escherichia coli* and *Serratia marcescens* ($\varepsilon'_{(-)} \approx 100$ Gram-negative). From each strain the same cell count and volume was measured. Particularly at a frequency of 50 MHz (maximum of discrimination), the real part of permittivity ε' of both Gram-positive strains was about 60 units higher than of the Gram-negative strains (Fig. 21)

The clear discrimination between the Gram-positive strains *Micrococcus luteus* and *Bacillus subtilis* as well as the Gram-negative strains *Escherichia coli* and *Serratia marcescens* at a frequency up to 100 MHz can be attributed to the β-dispersion. At these frequencies, proteins and other macromolecules of the bacterial cells polarize according to Markx *et al.* [35]. This effect decreases at frequencies above 100 MHz. With increasing frequency the influence of water becomes more prominent.

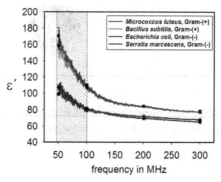

Figure 21. Discrimination of Gram-positive and –negative bacterial strains *via* dielectric spectroscopy. The diagram shows the real part of the permittivity ε' of the biomass of Gram-positive bacterial strains (*Micrococcus luteus* and *Bacillus subtilis* [upper curves]) and Gram-negative bacterial strains (*Escherichia coli* and *Serratia marcescens* [lower curves]) in a frequency range between 50 and 300 MHz. The highlighted area shows the most obvious region of differentiation between Gram-positive and Gram-negative bacterial strains [54].

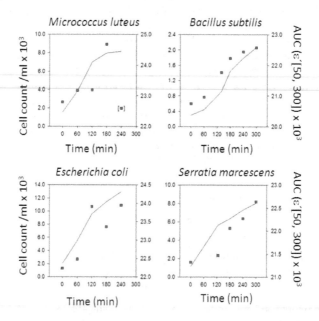

Figure 22. Monitoring of growth kinetics of four bacterial strains (growth phase). Red squares show the area under the curve (AUC) of the real part of permittivity ε' in a frequency range between 50 MHz and 100 MHz derived from measurements during the bacterial growth phase. The permittivity of the cell suspension was taken hourly for 240 or 300 min. Black lines show the cell count per ml taken at the same time as permittivity was measured [54].

All bacterial strains investigated in the present study revealed a characteristic time-dependent correlation between cell counts (black lines in Fig. 22) and ε' (red dots in Fig. 22). The growth kinetics was not influenced by the presence of accumulated metabolites in the culture medium since supernatants (liquid above precipitate) of every bacterial culture showed the same permittivity as the Standard I culture media (Fig. 22; $\varepsilon' = 85-78$ and $\varepsilon'' = 600-100$ @ $50-300$ MHz) [54].

3.4. Temperature influence on tissue permittivity

In therapeutic or diagnostic applications or biological effects of the electromagnetic field, dosimetric evaluations are greatly dependent on the precise knowledge of the dielectric parameters of biological tissues (relative permittivity ε and electrical conductivity σ). These parameters are sensitive to many influencing factors, which include the temperature of the target organ [55]. During radio-frequency or microwave radiation exposure, the internal temperature of tissue can change, thus influencing the electrical field distribution. For example, the evaluation of the lesion obtained by thermal ablation is a function of the relative permittivity and conductivity at 37°C and also of their evolution during heating. The influence of temperature in dielectric spectroscopy has been studied by several authors [56]-[58]. However, these effects remain misunderstood and the measured values are sparse at various frequencies and exist only for some organs.

To find out in how far temperature-dependent changes in permittivity can result in a parameter identified by ultra-wideband technology, water and different tissues were examined. To assess the basic capability of UWB radar for monitoring local temperatures, dedicated phantom and *in vivo* experiments were performed. Dielectric spectroscopy of water at different temperatures (25 – 80°C in steps of 5°C) and corresponding experiments using porcine and bovine tissue, such as udder, liver, muscle, and kidney revealed a distinct decrease of permittivity with increasing temperature. Nevertheless, heating of tissues to more than 60 °C might also reduce permittivity due to the reduction of water content. No distinct organ-specific differences in the temperature-dependent dielectric properties have been found so far (Fig. 23). Only fat, as low-water-content tissue, exhibited no influence on permittivity at different temperatures [59].

In addition to further studies with improved probes, corresponding analysis were performed using clinically approved temperature-based methods for tumor eradication, such as radio frequency ablation (RFA) or magnetic thermo ablation. For this experiment, a bovine liver was positioned onto a neutral electrode. The second, active electrode was launched into the liver tissue. Both electrodes together create a stress field, and the tissue around the active electrode becomes heated up to 60°C. Bi-static UWB antennas were first positioned in a distance to the region where RFA was thought to detect the signals of liver tissue itself. Then, the antennas were positioned above the region of radio frequency ablation, and changes in impulse response before, while and after radio frequency ablation were detected. The signal analysis displayed an increase of the impulse response during radio frequency ablation (data not shown) [59].

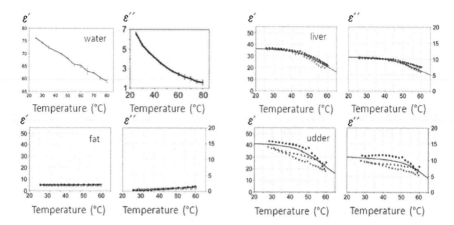

Figure 23. Real part (ε') and imaginary part (ε'') of the permittivity of water, liver, fat, and udder tissue at a frequency of 2 GHz is depicted. Water showed a constant decrease of permittivity in relation to the increase of the temperature. Liver and udder tissue showed a diminished deviation, and in contrast fat showed no change of the permittivity [59] at all.

The applicability of UWB for temperature monitoring was also assessed *in vivo* in mice. Prior to the start of experiments, mice were shaved at the abdominal region. Dielectric spectroscopy of the skin at the animal's liver region before and after euthanasia showed a time-dependent increase of permittivity as a result of decreasing temperature with on-going time after euthanasia. The data provide a good basis for further development of UWB as a non-invasive temperature measurement technology.

3.5. Permittivity variations by contrast media

Microwave-frequency dielectric contrast between malignant and normal tissue in the breast serves as the physical basis for emerging microwave methods of detecting and treating breast cancer. The effective dielectric properties of breast tissue are influenced at microwave frequencies by endogenous polar molecules, such as free and bound water, peptides, and proteins. Consequently, the dielectric properties depend on the type and physiological state of the tissue. The effective dielectric properties - both the dielectric constant and effective conductivity - can also be influenced by exogenous molecules introduced as contrast agents [60].

Detection of dielectric properties of structures and tissues with similar characteristics (such as breast and breast tumor) pose challenges for imaging by ultra-wideband technologies. Therefore, a phantom serving as a model for blood vessels with a constant flow of ethanol (infusion fluid) was created (Fig. 24 left panel) for first trials to test the sensitivity of the measurement apparatus. Additions of contrast agents (in this case a mixture of ethanol and water) were determined [25]. Such basic search is useful for finding suitable contrast agents including feasibilities and limitations regarding the detectability of, for example, concentration variations. The practice of clinical diagnostic radiology has been made possible by advances not only in diagnostic equipment and investigative techniques, but also in the contrast media that permit the visualization of the details of the internal structure of organs, which would not be possible without them. .The remarkably high tolerance of modern contrast media has been achieved through successive developments in chemical pharmacological technology.

The phantom serving as a model for blood vessels with a constant flow of ethanol was arranged. In the first step, the signals of this ethanol flow were received. By using a syringe *via* three-way cock 3 ml of the selected contrast agent (mixture of ethanol and water) were added, and the relative signal change was detected. The results show that with a decrease of water the signals become weaker (Fig. 24, right panel).

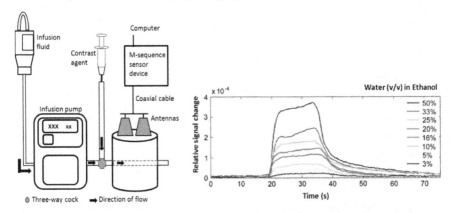

Figure 24. Depiction of the assembly of a phantom serving as a model for blood vessels with a constant flow of ethanol (left panel). The right panel depicts the relative signal variation (change of reflection coefficient) through adding 3 ml of contrast agent in relation to the virgin signal with constant flow of ethanol [43].

Even though dielectric spectroscopy of our group showed promising permittivity values of potential contrast agents such as physiological sodium chloride, the encountered permittivity increases *in vivo* are still to be enhanced to allow for a specific detection *via* UWB radar. One possibility to increase the capability of breast tumor imaging is the application of different clinically approved contrast agents such as ultrasound micro bubbles or iron oxide nanoparticles. Moreover, we expanded our investigations to the assessment of non-clinically approved agents (for example $BaSO_4$) in order to discover potential advantageous mechanistic conditions leading to local signal increase in terms of UWB diagnosis. Experiments will be systematically analyzed using dedicated phantoms, mimicking human tissues and blood flow.

Another challenge is the achievement of a selective accumulation of contrast agents in the target region to be detected by our UWB system. In this regard, a dynamic and transient accumulation *via* the tumor vascularization has been already postulated.

4. Remote organ motion tracking and its application in magnetic resonance imaging

4.1. Cardiac magnetic resonance imaging

Magnetic resonance imaging (MRI) is arguably the most innovative imaging modality in cardiology and neuroscience. It is based on the detection of precessing nuclear spins, mostly from protons of tissue water, in a strong static magnetic field. Using two additional kinds of magnetic fields, the position of the spins inside the human body can be encoded. To this end, the nuclear spin system is excited by resonant RF pulses at the precession frequency of the spin system. After excitation a macroscopic RF signal can be detected by an RF coil providing amplitude and phase information of the precessing nuclear magnetization. Applying additional magnetic field gradients the spin positions can be encoded by generating a well-defined spatial variation of the precession frequencies. Proper sequencing of spin excitation and gradient switching allows the reconstruction of 2D and 3D images from the acquired complex valued MR signals.

MRI data depend crucially on a multitude of physical parameters, e.g. moving spins will cause an additional phase modulation of the signal. One consequence is that MR images of the moving heart or of large vessels with pulsatile blood flow are severely distorted in the whole field of view. Hence, cardiac MRI (CMR) is seriously impaired by cardiac and respiratory motion when no proper gating with respect to both relevant motion types, cardiac and respiratory motion, is applied (Fig. 25). In clinically approved CMR procedures, electrocardiography (ECG) or pulse oximetry are used for cardiac gating and breath holding is applied for freezing respiratory motion [61],[62].

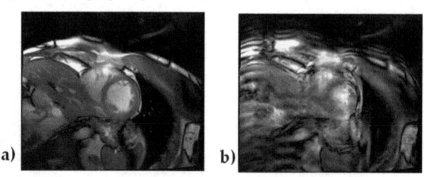

a)   b)

Figure 25. MR image (short axis view) of a human heart. a) Cardiac gating by pulse oximetry and breath hold; b) cardiac gating only, due to free breathing during image acquisition severe image artifacts occur

However, there are unmet needs of clinical CMR, particularly for high (≥ 3 T) and ultra-high (≥ 7 T) field MRI. Higher magnetic fields offer the chance to acquire images of better spatial resolution [63], but on the downside the ECG signal is increasingly perturbed by the magneto-hydrodynamic effect [64] until it becomes effectively useless for cardiac gating at

7 T. Furthermore, ECG electrodes are directly attached to the patient's skin, which may result in local RF burns. In addition, ECG and alternative approaches like pulse oximetric or acoustic cardiac triggering [65] do not provide any information about the respiratory state.

As a cardiac patient's breath hold is typically limited to about 15 s, a 3D whole heart coverage or imaging of the coronaries [66] would require proper respiration gating to acquire MR data under free breathing conditions. A well-established approach for respiration gating is the so-called MR navigator [67]. By means of some extra MR excitations, the momentary position of the diaphragm can be tracked over the respiratory. Unfortunately, these extra excitations interfere with the cardiac imaging sequence itself, making this technique complex and less reliable.

On this background, we propose the simultaneous use of multi-channel UWB radar and MRI to gain complementary information in particular for improving cardiac MRI. The anticipated potentials of this technique are (i) a contactless measuring principle for better patient safety and comfort, thus streamlining the clinical workflow, (ii) concurrent monitoring of a variety of body movements, (iii) direct relation to tissue mechanics [68], (iv) direct tracking of the temporal evolution of inner body landmarks, and (v) absence of any interferences of the UWB radar signals with the MR measurement if MR compatible designed UWB antennas are applied, (s. Section 2.3.3). The decomposition of physiological signatures in UWB radar data is the main challenge of this approach and a prerequisite for a reliable tracking of landmarks within the human body suitable for MRI gating.

Beyond MRI, there are a variety of other possible applications of in vivo UWB radar navigation systems in medical imaging or therapy. Examples are X-ray Computed Tomography (CT), Positron Emission Tomography (PET), Medical Ultrasonography (US), and Radiotherapy using photons or particles, or High Intensity Focused Ultrasound (HIFU). Lessons learned from all these approaches will foster medical applications of standalone UWB radar systems for intensive care monitoring, emergency medical aid, and home-based patient care [70].

4.2. Analytical and numerical modeling of the scenario

4.2.1. Channel-model

For modeling purposes, the human body can be approximated as a multilayered dielectric structure with characteristic reflection coefficients $\Gamma(f)$ (s. Fig. 26) [29], [71], [72]. The UWB signal, which can be a pulse or a pseudo-noise sequence [71] of up to 10 GHz bandwidth, is transmitted utilizing appropriate pulse-radiating antennas T_x (e.g., Double Ridged Horn or Vivaldi antennas). The reflected signal is detected by R_x, and the first step in further signal-processing usually is to calculate the correlation signal $R_{xy}(\tau)$ between received signal S_{Rx} and transmitted signal pulse S_{Tx} is [71], [72]. This represents the impulse response function (IRF) including the transfer functions of the antennas. By UWB measurement of the motion of a multi-layered dielectric phantom [29], the changes of reflections on the single interfaces can be found. Therefore, the signal variance $M(\tau)$ of the correlation signal $R_{xy}(\tau)$ is calculated.

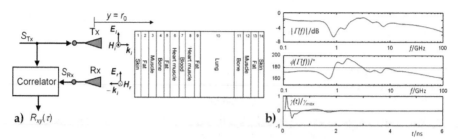

Figure 26. a) 14-layer arrangement to mimic the reflective properties of the human thorax (not to scale). E_i/E_r, H_i/H_r: incident/reflected electric/magnetic field component. k_i: wave vector of incident wave. **b)** Top: calculated magnitude of the reflection response $|\Gamma(f)|$, which is proportional to the frequency response function (FRF) of the object. Middle: unwrapped phase $\phi(f)$ of the reflection response $\Gamma(f)$. Bottom: normalized time domain representation $\gamma(t)$ of $\Gamma(f)$ impulse response function (IRF) of the reflection response.

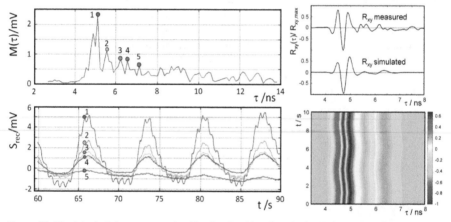

Figure 27. Physiological signatures received by the algorithm described in Ref. [29]. Top: Signal variance M(τ).Bottom: Physiological signatures corresponding to the label local maxima of M(τ).Top right: Measured and simulated correlation signal $R_{xy}(\tau)$.
Bottom right: Radargram of the measured and simulated correlation signal $R_{xy}(\tau)$.

The maxima in M(τ) correspond to the interfaces containing a considerable difference in the permittivity or are close to the illumination side if the transfer functions of the antennas are removed by de-convolution. By these maxima, the time signals corresponding to the interfaces can be extracted [29]. An example of simulated and measured correlation signal $R_{xy}(\tau)$ is given in Fig. 27, top right.

4.2.2. Analytical simulation of the intracranial pulsation detection

It is well known that simultaneously to the head's vibrations intracranial oscillations with spatial varying amplitude occur, induced by physiological sources [73]. Thus, it is only logical to ask whether these oscillations are detectable by UWB radar. Due to the simultaneous occurrence of the intracranial displacement and the vibration of the whole head, decomposing both signals requires sophisticated methods. As an initial step towards the solution to this problem, we need to get a feeling for the change in the acquired UWB reflection signal due to an intracranial oscillation. An analytical approach [71], [72] was applied which models the signal path and the oscillating stratified arrangement of the brain to get signals free of any interfering compositions. Figure 28 schematically depicts the set-up used to probe the human body with a UWB device, where S_{TX} symbolizes the excitation signal and S'_{TX} its temporal derivative representing the free space signal E_i in the channel. By the convolution of the impulse response function γ of the multilayered dielectric structure with S'_{TX}, the reflected electric field component $\gamma * S'_{TX} = E_r$ is archived and, accordingly, the received current signal $S_{RX} = (\gamma * S'_{TX})'$. The · symbol represents the convolution operator.

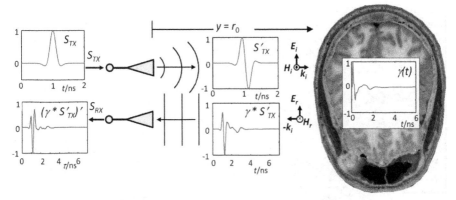

Figure 28. Signal path model for the current transfer function S_{RX}/S_{TX}.

Therefore, the spectral response of a dielectric medium is appropriately described in terms of a multiple Cole-Cole dispersion, which – by choosing parameters appropriate to each constituent - can be used to predict the dielectric behavior over the desired frequency range [71]. For such a layered arrangement, the reflection coefficient $\Gamma(\omega)$ can be calculated recursively. In this manner, the response of $\Gamma(\omega, t)$ to the variation of a certain internal interface can be analyzed [68], [73]. We simulated the physiological event by variations of $\Gamma(\omega, t)$, which is done by a sinusoidal oscillation of the white matter. Accordingly, the cerebro spinal fluid varies antipodally [76]. The correlation result $R_{xy}(\tau, t)$ was calculated just as its variation after a certain propagation time. The reconstruction of the intracranial motion applying the reconstruction algorithm proposed in [72] gave us a maximum deviation from the reference oscillation of about 4%. We conclude that the detection of intracranial oscillations using non-contact UWB is indeed feasible [72], [73]. It must be noted that for all real medical applications of this broadband technique trying to monitor variations of the body's interior, sophisticated signal processing techniques must be applied to decompose signals originating from the body's surface and signals originating from deeper sources [74]. The influence of the antenna's transfer function, in contrast, is less of an issue for real applications. For simplicity, we had assumed an ideal transfer function in the above simulation but non-ideal antenna behavior can be extracted from the received signal by using proper de-convolution techniques. Furthermore, the time courses of the ideal channel can be regained [72].

4.2.3. Full simulation of the electromagnetic field distribution

Beside the analytical approach, we are interested in the detailed temporal evolution of the electromagnetic fields inside and outside the human thorax. To this end, we investigated complex arrangements mimicking the illumination of a realistic human torso [75] model incorporating the geometry of the antennas by finite-difference time-domain method (FDTD) simulations. By FDTD simulation, we studied, e.g., the dependence of the illumination and detection angles of the transmission and receiving antennas on the quality of the received signal, *i.e.* the correlation result. In this way, an estimate of the optimized antenna placement can be found. Furthermore, by varying organs' boundaries by changing their thickness or/and placement of one or more tissue layers, different functional states can be investigated, e.g. the end-systolic and end-diastolic phase of the myocardium, which consequently determines a characteristic change of the received signals.

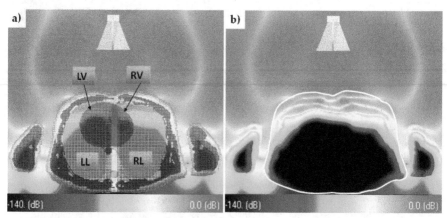

Figure 29. Extra- and intra-corporal electrical field distribution in an axial cross-section of the upper human body **a)** with tissue mesh; **b)** without the mesh showing the wave propagation intra-corporal. The thorax's contour is highlighted by the white line.

An example of the complex wave propagation inside the human torso is shown in Fig. 29. Due to the higher permittivity ε inside the body, the propagation velocity is slowed down according to $c = c_0/\sqrt{\varepsilon}$. Hence, a bending of the extra- and intra-corporal wave fronts results. The transmitted spherical wave fronts are refracted towards the center of the thorax, which is beneficial for the illumination of the myocardial section lying deeper inside the thorax. By these simulations, we achieve an in-depth understanding of the complex electromagnetic field distribution and the dependencies of the resulting output signal of the receiving antenna [73]. Therefore, the results of these simulations are helpful to increase the accuracy of reconstructed physiological signatures from deep sources by finding the optimized antenna position regarding the better penetrability of selected body areas. This, of course, requires the adaptation of the model to the actual thorax geometry of the patient as obtained by MRI scans.

4.3. Detection of motion by UWB radar

4.3.1. Motion detection for a multilayered phantom

We compared the motion detection by variance calculation in a combined MRI/UWB measurement using appropriate MR-compatible phantoms [29]. The dielectric phantoms were arranged in a sandwich structure to mimic the sequence of biological tissue layers of the human thorax.

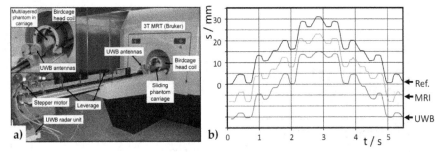

Figure 30. a) Set-up of the combined MRI/UWB measurement; **b)** Comparison of the reference profile with the data obtained simultaneously from MR and UWB radar measurements. The profiles are offset for clarity.

Such a sandwich was placed in a moveable sledge-like fixture inside a birdcage MR head coil. The motion profile of the sandwich structure was shaped to approximate respiratory motion of the thorax superimposed by cardiac oscillations (Fig. 30). An M-sequence UWB-Radar system (up to 5 GHz) [76] and MR compatible UWB antennas [10], [32] were utilized to detect the motion of the phantom inside a 3-T MR scanner (Bruker MEDSPEC 30/100). A flow-compensated gradient echo CINE (time resolution 50 ms) sequence was used to reduce artifacts generated by the phantom movements.

Additionally, the physiological signatures monitored by UWB-radar were validated by comparison to simultaneously acquired MR measurements on the same subject [13], [29], [77] (cf. Section 4.5.2 and 4.6).

4.3.2. Detection of micro motion

Subject motion appears to be a limiting factor in numerous MR imaging applications especially in the case of high and ultra-high fields, e.g. high-resolution functional MRI (fMRI). For head imaging the subject's ability to maintain the same head position is limiting the total acquisition time. This period typically does not exceed several minutes and may be considerably reduced in the case of pathologies. Several navigator techniques have been proposed to circumvent the subject motion problem [73]. MR navigators, however, do not only extend the scan because of the time necessary for acquiring the position information, but also require additional excitation pulses affecting the steady-state magnetization. Furthermore, if the very high spatial resolution offered by ultra-high-field MR scanners shall be exploited, the displacements caused by respiration and cardiac activity have to be considered. Thus, we propose to apply an UWB radar technique to monitor such micro motions.

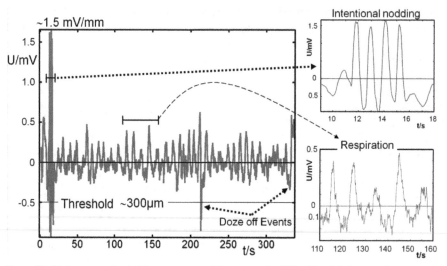

Figure 31. Motion reconstructed from a measured time interval of 350 s. The right inset at the top displays the four nodding events (~1 mm amplitude, episode [t = 10 s,..., t = 18 s]) to localize the surface of the head. Respiratory displacements are clearly visible (right inset bottom, episode [t = 110 s,..., t = 160 s]) and spontaneous twitches are highlighted.

First *in-vivo* motions reconstructed from a measured time interval of 350 s are shown in Fig. 31. By applying appropriate filters in a selected time interval even the cardiac induced displacements were detected with an amplitude of about 40 μm. Thus, we could detect all kinds of involuntary motions (respiratory, cardiac), but also doze-off-events are visible, demonstrating the feasibility of interfacing an MR scanner with an external UWB radar based motion tracking system. Our system is capable of determining the position of interest with sub-millimeter accuracy and an update rate of 44.2 Hz. Using the UWB tracking data of the volunteer's head, the motion artifacts can be compensated for in real time or by post-processing enhancing the actual resolution of the MR scan [73].

4.3.3. Separation of motion components by blind source separation

Monitoring the motion inside the human body, the correlation functions of transmitted and received signals (i.e. the IRF) contain a mixture of all simultaneously occurring motions.

Especially for the human torso where - due to higher harmonics from the highly nonlinear respiratory cycle - the separation of the cardiac cycle by common signal filtering in the frequency domain is limited, another separation of motion components is necessary. For this reason, the separation of motion components based on blind source separation (BSS) was developed.

The IRF from a single UWB shot is a time series of 511 data points with a dwell time of 112 ps. This defines an IRF time scale of 57 ns but is still instantaneous compared to anatomical motions. These shots are then repeated for instance 2000 times at a rate of 44 Hz covering a total time span of 45 s. For further analysis, only the most interesting regime of the IRF data is considered. These are the 100 data points, i.e. a window of 11.2 ns, right after the IRF maximum due to direct cross-talk between Tx and Rx antenna. Following the temporal evolution of each selected data point over the 2000 repetitions, 100 virtual channels are obtained and subjected to BSS decomposition (ROI, see Fig. 32.a). By removing the mean values in these virtual channels, the changes of the radar signal on the anatomical time scale become visible, see Fig. 32.b. The motion pattern is dominated by respiration; cardiac motion is considerably smaller and not immediately visible in the raw data.

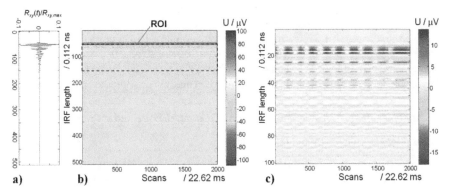

Figure 32. a) Single IRF and b) radargram of one channel with region of interest and c) selected 100 virtual channels, mean value removed

The data analysis is based on the BSS and assumes a measured signal $x(t)$ to be a linear combination of unknown zero-mean source signals $s(t)$ with an unknown mixing matrix $\mathbf{A}$:

$$x(t) = \mathbf{A}s(t) \quad x = (x_1,...,x_m)^T . \tag{2}$$

The original sources $s(t)$ can be estimated by the components $y(t)$ which can be calculated from the estimation of the de-mixing matrix $\mathbf{A}^* \approx \mathbf{A}^{-1}$:

$$y(t) = \mathbf{A}^* x(t) = \mathbf{A}^* \mathbf{A}s(t) \tag{3}$$

In our analysis, a second-order time-domain algorithm (TDSEP, Temporal Decorrelation source SEParation) was applied which is described in detail in [78]. In TDSEP the unknown mixing matrix $\mathbf{A}$ is calculated by simultaneous diagonalization of a set of correlation matrices $\mathbf{R}_\tau(x)$ for different choices of τ.

$$\mathbf{R}_{\tau(x)} = \left\langle x(t)x^T(t-\tau) \right\rangle$$

$$\mathbf{R}_{\tau(x)} = \left\langle \mathbf{A}s(t)(\mathbf{A}s(t-\tau))^T \right\rangle = \mathbf{A}\mathbf{R}_{\tau(s)}\mathbf{A}^T \tag{4}$$

where the angular brackets denote time averaging. The quality of signal separation depends strongly on the choice of τ. However, solving $\mathbf{R}_\tau(x) = \mathbf{A}\mathbf{R}_\tau(s)$ $\mathbf{A}^T$ for several τ by simultaneous diagonalization eliminates this obstacle. It is recommended by biomagnetic research to choose the number of time shifts τ larger than 40 and to include the time constant of those components which are known a priori, e.g. the range of possible cardiac frequencies $1/\tau_{cardiac}$ [79]. Additionally, Principal-Component Analysis (PCA) compression was applied to reduce the number of channels used for generating the correlation matrices $\mathbf{R}_\tau(x)$ and reduce computation time for the BSS. The components of the resulting sources are calculated using eq. (3). Automatic identification of the cardiac component was provided by a frequency-domain selection criterion because for non-pathological conditions the main spectral power density of the heart motion falls in a frequency range of 0.5 Hz to 7 Hz. The algorithm searches for the highest ratio between a single narrowband signal (fundamental mode and first harmonic) within this frequency range and the maximum signal outside this range. A high-order zero-phase digital band pass filter of 0.5–5 Hz was applied to the cardiac component of the UWB signal. In a similar way, respiration can be identified by the BSS component with the maximum L2 norm in the frequency range of 0.05 Hz to 0.5 Hz.

4.4. Analyses of cardiac mechanics by multi-channel UWB radar

4.4.1. Compatibility of MRI and UWB radar

Compatibility is the most challenging issue when combining MRI with other modalities. Therefore, the UWB antennas employed are important parts. Eddy currents due to the switching magnetic gradient fields as well as the interference with the powerful RF pulses from the MRI scanner, see Section 2.3.3, were minimized by proper antenna design. The cut-off frequency of the MR-compatible double ridged horn antennas at 1.5 GHz [32] marks the lower limit of the transmitted and received signal frequencies. Coupling to the narrowband MRI frequencies (300 MHz at 7 T, 125MHz at 3T) is thus minimized in both directions. Additionally, the inputs of both our UWB radar systems (MEODAT GmbH, Ilmenau, Germany), one single-module: 1Tx-2Rx-device and one four-module multi-input-multi-output: 4Tx-8Rx-MIMO device, were protected by 1.2 GHz high pass filters. In both UWB systems, the transmitted radar signals were generated by a pseudo-random M-sequence. With $m = 9$ it has a length of $2^m-1 = 511$ clock signals at $f_0 = 8.95$ GHz [76]. The equivalent UWB power spectrum extends up to $f_0/2$. The impulse response function (IRF) is obtained as mentioned before by correlation of the received signal of the investigated object with the M-sequence [76]. By means of this technique, the signal-to-noise ratio is improved due to the removal of the uncorrelated noise by the correlation of the received signals with the transmitted signal pattern. In this way, even smallest parts of the RF pulses of the MRI were avoided.

4.4.2. Multi-channel UWB radar applying two receiver channels for cardiac trigger events

We started our multi-channel UWB radar development with the single module device enabling us to add a second receiver (*Rx*) antenna, oriented towards the left-anterior oblique direction [68] (Fig. 33.a), to the existing *Rx* and *Tx* antennas facing the antero-posterior direction. The UWB data were recorded at 44.2 Hz. Corresponding to the data selection in Section 4.3.3, we obtained 200 virtual data channels from the IRFs of two UWB measurement channels for the decomposition by blind source separation (BSS).

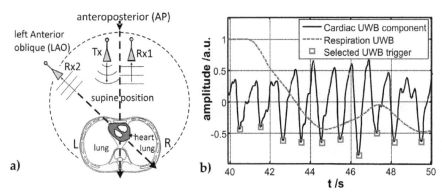

Figure 33. a) Scheme of the UWB radar with one transmitter (*Tx*) and two receiver (*Rx*) antennas and measurement set-up; **b)** Cardiac UWB signal applying both *Rx* channels and the calculated trigger events in the signal by combination of low peak and slew rate calculation.

In the cardiac UWB signal, we chose the points of maximum myocardial contraction during the heart cycle. These points are related to the minima of the UWB signal (Fig. 33.b: squares). To increase the robustness of this detection scheme, we combined it with a simple difference calculation at the trailing edge of the minima. Additional consistency checks on the oscillation amplitude were used to suppress double triggering.

By employing two *Rx* channels (Fig. 33.a) the UWB radar detection of the cardiac cycle worked reliably, even in the free breathing mode. In simple cases, e.g. under breath-holding conditions, it is possible to detect cardiac motion with just one *Rx* channel. However, this will not work in general, more complicated situations.

4.4.3. Application of up to 32 receiver channels

By using two *Rx* channels, it was still necessary to align the antennas properly towards the heart. This becomes more critical for measurements during cardiac MRI where the MR coil is placed on the chest of the subject, partly blocking the free line of sight between radar antenna and the heart, see Fig. 38. With our development of a multiple-antenna set-up it is much easier to handle this adjustment by just choosing the 'good channels' in a pool of available channels.

By integration of a MIMO UWB device (MEODAT GmbH, Ilmenau, Germany) containing four modules, each with one Tx and two Rx channels [76], up to 32 channels became available. In a 1 $Tx * 8 Rx$ configuration a sampling frequency up to 530.4 Hz can be realized. Using sequentially activated transmitters the set-up can be extended to 32 channels (4 $Tx * 8 Rx$) at a reduced sampling frequency of up to 132.6 Hz. For cardiac motion detection, the four Tx and eight RX antennas are placed over the chest as depicted in Fig. 34 and adjusted to aim for one central point at a distance of 100 cm.

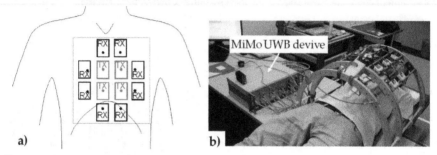

Figure 34. a) Scheme of the UWB radar set-up with 8 Rx and 4 Tx antennas b) MR compatible measurement set-up.

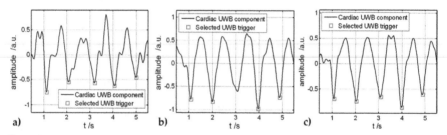

Figure 35. Cardiac signal and detected trigger events for **a)** two hand-picked best channels, **b)** all 32 channels, **c)** the 24 "good channels".

The procedure to identify the most useful channels for triggering starts with a short preparatory measurement, where each channel is analyzed by the BSS to decompose the complex UWB signals [80], extracts the relevant cardiac component and calculates the trigger events as described in Section 4.4.2. The quality of each measurement channel is assessed by calculating the variation of the time interval between trigger events. For comparison, Fig. 35.a depicts the result of the BSS analysis by the best two channels, manually selected for the smallest variation between the trigger events. The cardiac signal based on these two hand-picked channels represents the best achievable result for a set-up like in the section before. By utilizing all 32 channels for the BSS, a smoother cardiac signal is detected, and the motion amplitude shows less variation over the time. However, the sharpness of the trailing slopes is also reduced. Due to this fact, the third trigger event escaped detection (s. Fig. 35.b). Some of the 32 channels contained much noise resulting in a jitter of their trigger contributions and smearing out the sharpness in the combined signal determined by the BSS. By the preparatory check those channels with the highest variation in their 'cardiac' signals were excluded as they were likely contaminated with noise or other motion components. By rejecting the eight noisiest channels and recalculating the BSS with the remaining channels, a cardiac signal is obtained with sharp trailing slopes and well-defined trigger events (s. Fig. 35.c).

The primary goal of this development was to simplify the system handling during cardiac navigation for high-resolution MRI. In addition, the capability of monitoring non-invasively the cardiac activity of a person in an unknown position, e.g. in a patient bed, can be important for a variety of novel medical applications in clinical medicine and biomedical research. As multi-channel UWB radar is unimpeded by bedding or clothing, it is applicable not only in conjunction with MRI. It would also be a valuable stand-alone modality for intensive care monitoring of patient groups not permitting the use of skin contact sensors. Neonates, children at risk of sudden infant death syndrome or burn victims are just a few examples.

4.4.4. Illumination of human thorax by multiple antenna groups

Stand-alone UWB radar enables the detection of cardiac activities by different illumination conditions as shown in [68] for the radiographic standard position. The illumination of the heart from only one side at a time, like the frontal direction for motion detection as depicted in Fig. 33, was extended to the simultaneous illumination of two sides. No averaging was performed to enable the comparison of single heart beats [30]. This approach can open the field for new diagnostic applications by detecting differences and disturbances in comparative measurements of the left and right ventricle, thus recognizing potentially pathological patterns [69]. Two groups of four Rx and one Tx antennas were applied for this purpose. The first was placed in the left lateral and the second in the right anterior oblique position.

Each antenna group consisted of a single Tx antenna surrounded by four Rx antennas. All antennas were directed towards the estimated center position of the heart. The challenge was to measure the cardiac motion even from the lateral position, where the attenuation of the reflected signals from the heart is much higher due to the prolonged propagation path in tissue. The data analysis by BSS was applied for both antenna groups separately. For comparison, the data of only two or all four Rx channel per group were analyzed.

For lateral position, the UWB signal from the cardiac motion is considerably weaker and much more affected by noise. However, by increasing the number of Rx- channels, the signal quality improved substantially, effectively compensating the strong signal attenuation (s. Fig. 36.b). Only healthy volunteers were examined in this particular study but even among them characteristic peculiarities can be found. In both ventricles, the contraction velocity (trailing edge of the UWB motion curve) is higher than the velocity of ventricle dilatation. The duration of the dilation period, on the other hand, is longer for the right ventricle compared to its counterpart on the left. More characteristic features are expected to be visible in patients with cardiac diseases or malfunctions.

4.5. Simultaneous cardiac UWB/ECG, UWB/MRI measurements

4.5.1. UWB radar and high resolution ECG

UWB and ECG were simultaneously acquired. The radar system was equivalent to Section 4.4.2 with one Tx and two Rx channels. The ECG was recorded with two channels (left arm and left leg against right arm) at a sampling frequency of 8 kHz. For the UWB signals sampled at 44.2 Hz, the same data analysis (see Section 4.4.2) was applied to extract the cardiac signal and determine the trigger events. The usual R-peak detection was applied to trigger on the ECG signal. Cardiac UWB and ECG signals were both re-sampled at 1 kHz to retain more detailed information of the ECG.

The point of maximum mechanical contraction of the heart in the cardiac UWB signals (s. Fig. 37.a) is delayed to the ECG R-peak, indicating the point of the myocardium's peak electrical activity. Therefore, we have to be aware of the difference between detecting cardiac mechanics by UWB radar and the heart's electrical activity by ECG. For the goal of MRI gating, however, the important thing is the existence of a fixed temporal relationship between ECG and UWB signals with as little jitter as possible. For the time lag between ECG and UWB trigger events, we obtained a standard deviation of less than 20 ms which is already smaller than the UWB sampling time of 22.6 ms. This result proves the consistency and robustness of our procedure.

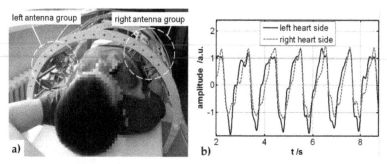

Figure 36. a) Measurement set-up with two antenna groups for separate monitoring of the heart's left and right ventricle; b) Cardiac signal for left and right ventricle applying four Rx channels.

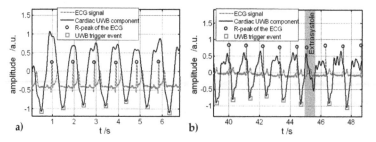

Figure 37. a) ECG signal with R-peak trigger events and UWB signal with trigger events located at the maximum of mechanical contraction; b) Measurement with an extra-systole.

Another measurement example depicted in Fig. 37.b shows a cardiac sequence containing an extra-systole at 45.5 s. In the ECG signal, this appears as a spontaneous change in the R-R-duration. In the cardiac UWB signal, we observe an incomplete contraction of the myocardium due to the "erroneous" electric excitation. Consequently, no trigger event was generated by this extra-systole. This ability to analyze cardiac mechanics by stand-alone UWB radar or in conjunction with ECG can be exploited, e.g., for infarction detection, as ischemic tissue shows a modified contraction pattern.

4.5.2. Comparison of cardiac UWB Signal and one dimensional MRI

For better understanding the relationship between actual cardiac mechanics and UWB motion-detection signals, a fast MR-sequence was developed with the aim to monitor myocardial landmarks inside the human body in real time. We implemented a very fast 1D gradient echo sequence for low RF power deposition in tissue and high scan repetition frequency on our MR scanner [77]. One dimensional MR profiles and motion sensitive UWB data were acquired simultaneously allowing the comparison of both techniques and hence a verification of the UWB radar navigator. MR compatible UWB antennas [32] mounted above the chest were directed towards the heart (s. Fig. 38). A flexible RF coil with large openings was used to detect the MRI signal. The UWB data were sampled at 132.6 Hz. Using one Tx and five Rx UWB antennas 500 virtual channels could be constructed from the IRFs.

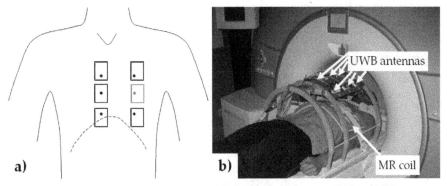

Figure 38. a) Scheme of the antenna configuration; b) Set-up of simultaneous UWB and MRI measurement.

In the MRI sequence, the one-dimensional 'pencil-like' imaging region is selected by the intersecting volume of two perpendicular slices (s. Fig. 39.a). Both slices are excited in short succession resulting in a saturation effect in the region of the intersection. When the experiment is repeated with a different delay time between both excitation pulses, the two images differ only in the strength of this saturation effect, and subtraction yields the desired 1D image. Placed through the heart in antero-posterior direction, this 'pencil' was scanned at a repetition frequency of 25.4 Hz. The motion components in both data sets, the 500 virtual UWB channels and the MR pencil, were once again separated by applying BSS decomposition.

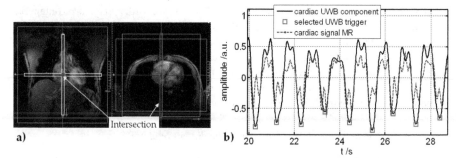

Figure 39. a) Selection of the 'pencil' by two crossing slices in antero-posterior direction through the heart; **b)** Detected cardiac motion component by UWB radar and MR "pencil".

The trigger events (squares in Fig. 39.b) in the UWB cardiac motion data - representing the point of maximum contraction of the myocardium - were determined by applying the algorithm proposed by us. This did not work with the cardiac components of the MR signal due to the pronounced double peaks in this data set. Comparing the cardiac components simultaneously gained by UWB and MR data, we observe perfectly matching slopes of both signals. However, in contrast to UWB radar the MR signal is affected by the blood velocity in the heart producing the double peaks. Keeping this in mind, we can conclude that both modalities render the same motion. Thus, we can assume the cardiac motion detection by UWB radar to be verified.

4.6. Advances for respiratory motions

To establish a UWB navigator for cardiac MRI in free-breathing mode, a landmark tracking of the heart due to the respiration motion is necessary [66]. The time-dependent UWB radar signal contains mainly respiratory motion of the chest, as in Fig. 32.b, which is not necessarily identical to the mechanical displacement of the heart. Therefore, we compared the UWB detected respiratory motion to simultaneously measured 1D MRI as described in Section 4.5.2. Resulting from that comparison, we extended the UWB configuration to allow for the detection of abdominal respiration, too (Fig. 40.a). The antenna configuration applied in the comparison made in Section 4.5.2 was extended by two additional channels above the chest and a second group with one Tx and one Rx antenna over the abdominal region.

In MR-based navigator techniques [67], the position of the diaphragm is monitored because the shift of the diaphragm is the dominant motion component of the heart due to respiration. The displacement of the diaphragm is mainly orientated in head-foot direction. Hence, the pencil-like one dimensional MRI was placed in head-foot direction across the heart. The UWB data of the first antenna group were decomposed by BSS for detection of breast respiration and cardiac cycle and the second group for the abdominal respiration. In the same way, the motion components were decomposed for "pencil-like" MRI.

The UWB detected breast respiration is not suitable to monitor the mechanical heart shift in head-foot direction. In Fig. 41, a delay between the breast respiration and the heart shift is depicted, and especially in Fig. 41.b it becomes obvious that these are different processes. However, the UWB detected abdominal respiration correlates well with heart motion due to respiration. The correlation factor in measurement a) is 0.932 and 0.81 in measurement b).

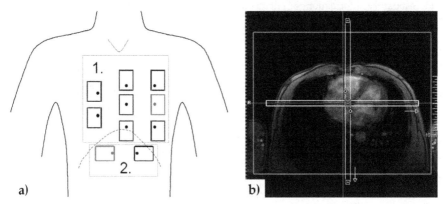

Figure 40. a) Extended antenna configuration with second group over the abdominal region; **b)** Placement of the two slices for the 'pencil-like' MRI (Head ⇔ Foot).

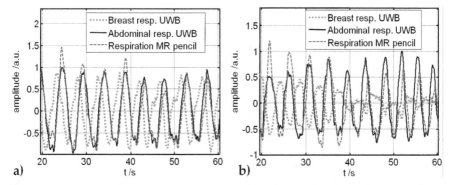

Figure 41. Breast and abdominal respiration by UWB radar and mechanical heart shift in head ⇔ foot direction monitored by MR pencil **a)** with changed breast and abdominal respiration **b)** with fading breast but changed abdominal component.

4.7. UWB triggered cardiac MRI

CMR and UWB signals were acquired simultaneously and synchronously to enable UWB triggering [81]. The UWB antennas were mounted in the same frontal position related to the subject as in Section 4.4.1. Simultaneous pulse oximetry (PO) was applied to compare our approach with another established triggering technique for cardiac MRI.

After acquiring a series of CMR images using a clinical sequence with conventional PO gating, we retrospectively reconstructed the k-space data a second time but now using trigger points derived from the simultaneously acquired UWB radar signals [81]. Figure 42.b shows that both methods give virtually undistinguishable results, thus establishing the feasibility of CMR imaging utilizing non-contact UWB radar for triggering. In contrast to established techniques like ECG or PO, however, contact-less UWB-sensing provides cardiac and respiratory information simultaneously and, thus, a sequence-independent external navigator signal.

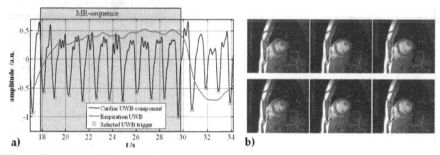

Figure 42. a) Cardiac UWB signal with selected trigger events; **b)** Top: Reconstructed images utilizing PO trigger, Bottom: Image reconstruction by UWB trigger events applied.

5. Microwave imaging in medicine

5.1. Introduction

Microwave ultra-wideband (UWB) sensing and imaging represents a promising alternative for the early-stage screening diagnostics of breast cancer. This perspective results from advantageous properties of microwaves: sensitivity of the dielectric properties of human tissue to physiological signatures of clinical interest in this frequency range, especially water content, their non-ionizing nature (compared to X-rays), and the potential of a cost-efficient imaging technology (compared to MRI) [82].

Numerous research groups have been working in this field since the end of the 1990s. Many studies deal with simulations, several groups perform phantom measurements, but only very few have already started some first clinical measurements. The challenges which have to be met concerning real *in vivo* measurements are multifaceted and depend on the conditions of the measurement scenario. The developed strategies and measurement principles of microwave breast imaging can be classified according to various characteristics: active vs. passive vs. heterogeneous microwave imaging systems [83]; microwave tomography (or spectroscopy) imaging [84] vs. UWB radar imaging [85]; examination in prone vs. supine position [83] and some further differentiations. This chapter deals exclusively with active microwave imaging based on the UWB radar principle which can be applied in general in both examination positions.

Figure 43 shows two basic antenna arrangements for the prone examination position. They differ in the antenna-skin distance.

Non-contact breast imaging: The most significant reason for non-contact breast measurements is the size of the antennas compared with the breast size. Thereby, it is impossible to mount a sufficient number of antennas on the breast surface in order to achieve an adequate image quality. The displacement of the antennas from the breast increases the area where additional antennas can be localized. Besides that, it allows mechanical scanning where the antennas can be rotated around the breast in order to create a synthetic aperture. On the other hand, this non-contact strategy is accompanied by a lot of other problems and challenges.

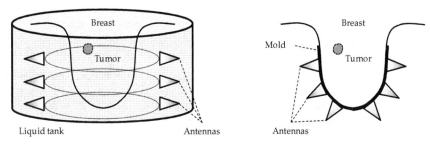

Figure 43. Schematization of non-contact breast imaging using a liquid contact medium (left side) and contact-based breast imaging (right side) in the prone examination position

Depending on the dielectric contrast between the medium surrounding the antennas and the breast tissue, only a fraction of the radiated signal energy will penetrate the breast. The major part will be reflected at the breast surface. It provides clutter which has to be eliminated since it perturbs the signals of interest. In order to reduce the reflection coefficient, several approaches use a liquid coupling medium in which the breast has to be immersed and in which the antennas can surround the breast. The same energy reduction effect appears for reflected components from inside of the breast passing the dielectric boundary in the opposite direction. Furthermore, in the opposite direction (from dielectric dense medium into a less dense medium) waves can only leave the breast below the angle of total reflection which implies an additional reduction of the detectable signal energy outside the breast.

The individual breast shape plays an important role in connection with these effects as well as for image processing. In section 5.3, we describe a method for breast and whole body surface reconstruction based on the reflected UWB signals.

Contact-mode breast imaging: Contact-based breast imaging avoids the disadvantages described above. The antennas are localized directly at the breast surface. Understandably, they have to be small enough in order to arrange a sufficient number of antennas around the breast. The corresponding number of signal channels will be obtained by electronic scanning, that means sequential feeding of all transmitter antennas with simultaneous signal acquisition of all receiving antennas. This strategy involves the problem of individual breast shapes and sizes which influences the contact pressure of the breast skin onto the antenna aperture and, thus, the signal quality [86].

However, we prefer this measurement scenario for our current investigations, and intend to weaken the contact problem in the future by 2 or 3 different array sizes and an additional gentle suction of the breast into the antenna array by a slight underpressure. In section 5.4, we present an experimental measuring set-up where we pursue a strategy of nearly direct contact imaging in order to conjoin the advantages of contact-based imaging with the possibility of mechanical scanning.

5.2. Breast and body phantoms

In the context of UWB tissue sensing, the water content plays a key role as it determines the inherent dielectric properties (ε' and ε'') [43]. Moreover, the water content is known to vary among the different human tissues as well as between specific normal and pathologic ones, thus offering a potentially broad spectrum of UWB applications for biomedical diagnostics.

Oil-in-gelatin phantoms, mimicking the dielectric properties of human tissues, were manufactured according to a protocol from [24]. The water concentration varied between 19 and 95% (v/v; ~ 10% water graduation steps), to obtain a set of materials with different permittivity values (ε' ranging from 8 to 59 and ε'' ranging from 0.5 to 11, both averaged over frequencies from 1 to 4 GHz). The measurements were carried out by means of the M-sequence devices [76], [87] with HaLoS chipsets and a frequency bandwidth of 4.5 GHz, as well as the radar data acquisition and analysis software "ultraANALYSER" developed for this purpose.

The variation of the oil-water-concentration led to the identification of distinct permittivity values ε' (Fig. 44, insert) of the different oil-in-gelatin phantoms. The phantom, which was manufactured without oil (95% water), showed values between 53 and 59 for the real part ε' and between 11 and 10 for the imaginary part ε'' of the permittivity in the frequency range between 1 and 3.5 GHz (Fig. 44, insert). The results for pure distilled water are also displayed. The real part of permittivity agrees well with literature data [88].

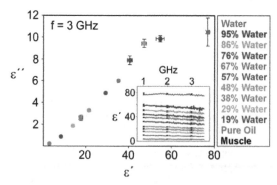

Figure 44. Dielectric properties of nine oil-in-gelatin phantoms with varying percentage of water (from 19% to 95% water (v/v)) and porcine muscle tissue. Depicted is the correlation of the real part ε' and the imaginary part ε'' of the phantoms. Both parts increase with an increasing water-concentration. Error bars represent the standard deviation from an average of three individual measurements on the phantom surface (insert).

5.3. Breast and body surface reconstruction

5.3.1. Method

The benefits of the exact knowledge of the breast surface for non-contact microwave breast imaging are manifold and can improve the results significantly. The inclusion of the breast shape information is essential to calculate the wave traveling path in order to image the interior of the breast based on radar beam-forming techniques. Some approaches use the surface information for initial estimations. Other non-contact measurement approaches strive to illuminate the breast from a specific distance which requires a very fast online surface identification in order to adapt the antenna position during measurement. Furthermore, in the case of varying distances between antenna and breast, the exact knowledge of the breast surface can improve the estimation of the skin reflection component for a better early time artifact removal. In order to reduce the calculation time, the region of interest (i.e. the region for which the image has to be processed) can be restricted based on known surface geometry [89], [90].

Additionally to the significance for breast imaging, UWB microwave radar is suitable for whole body surface reconstruction which can be used in other medical microwave applications as well as in safety-relevant tasks, e.g. under-dress weapon detection.

The Boundary Scattering Transform (BST) represents a powerful approach for surface detection problems. BST and its inverse transform (IBST) were introduced 2004 by Sakamoto and Sato [91] as basic algorithms for high-speed ultra wideband imaging, called SEABED (Shape Estimation Algorithm based on BST and Extraction of Directly scattered waves). Since then, this idea has been extended from mono-static 2D-imaging to the point of bi-static 3D-imaging (IBBST) [92]. The SEABED algorithm represents a high–speed, high-resolution microwave imaging procedure. It does not include the entire radar signal; it uses only wave fronts instead. Furthermore, changes (derivatives) of the propagation time (transmitter → object surface → receiver) depending on the antenna position during the scan process play an important role. SEABED consists of three steps: 1. Detection of the wave fronts and calculation of their derivatives with respect to the coordinates of the scan plane. 2. Inverse Boundary Scattering Transform, which yields spatially distributed points representing the surface of the object. 3. Reconstruction of the surface based on these points.

The practical applicability of the original algorithm to the identification of complex shaped surfaces is limited because of the inherent planar scanning scheme and, therefore, the disadvantage of illuminating only one side of the object. For this reason, we extended the bi-static approach of [92] toward non-planar scanning and a fully three-dimensional antenna movement based on the idea that in the case of arbitrary non-planar scan schemes the current scan plane can be approximated by the tangential plane at each antenna position [93]. An antenna position dependent coordinate transform which ensures that the antenna axis is parallel to the x-axis and the current scan plane is parallel to one plane of the coordinate system allows the application of the IBBST for nearly arbitrary scan surfaces. More precisely, this generalized approach is limited to scenarios where the antennas will be moved orthogonally or parallel to the antenna axis, which is fulfilled in most practical cases. First results of breast shape identification were published in [94], [95].

Based on the following transform equation, the coordinate of the specular point can be calculated

$$\bar{x} = \bar{X} - \frac{2D^3 D_{\bar{X}}}{D^2 - d^2 + \sqrt{(D^2 - d^2)^2 + 4d^2 D^2 D_{\bar{X}}^2}}$$

$$\bar{y} = \bar{Y} + \frac{D_{\bar{Y}}}{D^3}\left(d^2(\bar{x} - \bar{X})^2 - D^4\right) \qquad (5)$$

$$\bar{z} = \bar{Z} + \sqrt{D^2 - d^2 - (\bar{y} - \bar{Y})^2 - \frac{(D^2 - d^2)(\bar{x} - \bar{X})^2}{D^2}}$$

where $\bar{x}, \bar{y}, \bar{z}$ are the coordinates of the reflective surface point (specular point), $\bar{X}, \bar{Y}, \bar{Z}$ are the coordinates of the center between the two antennas, D is the half distance transmitter → reflection point → receiver, d is the half distance between the two antennas, and $D_{\bar{X}} = \frac{dD}{d\bar{X}}$, $D_{\bar{Y}} = \frac{dD}{d\bar{Y}}$ symbolizes the derivatives of the distance with respect to the denoted direction of antenna movement. The bars above the symbols mark the coordinates of the transformed coordinate system [93].

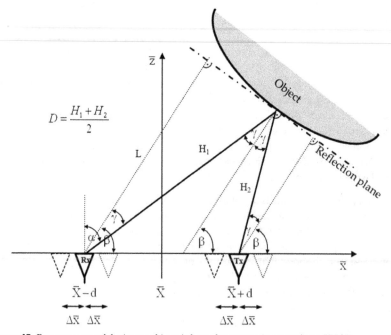

Figure 45. Ray geometry of the inverse bi-static boundary scattering transform (IBBST)

The main challenge is the exact detection of the wave fronts and their proper derivative. For the purpose of wave front detection, we use an iterative correlation-based detection algorithm similar to [96]. In this connection, a short antenna impulse response over a sufficiently wide angular range plays an important role. The difficulties of obtaining appropriate wave front derivatives result from the three-dimensional nature of the problem. The antennas are moved and the transmitted waves are reflected in the three-dimensional space. Especially in the case of wave front crossing and impulse overlapping as well as sparsely detected wave fronts, it is very complicated to recognize which identified wave front at one scan position is related to which wave front at the previous scan position and vice versa. So, it may happen that derivative values are wrongly calculated, which can lead to a spatially false projection of the surface points. In order to avoid such errors, we establish thresholds of feasible derivative values dependent on the antenna beam width.

5.3.2. Detection and elimination of improper wavefronts

General limit values: The range of values of the distance derivatives $D_{\bar{x},\bar{y}}$ is theoretically bounded between 0 and 1 depending on the slope of the reflection plane (tangent plane of the object surface at the specular point). In the case of parallelism between reflection plane and antenna axis, $D_{\bar{x}} = 0$, whereas in the case of orthogonality, $D_{\bar{x}} = \pm 1$. Thus, calculated values $|D_{\bar{x}}| > 1$ are definitely caused by incorrect wave front detection. Consideration of these general boundaries and exclusion of wave fronts exceeding them yields a significant improvement.

Customized plausibility limit values: The boundary $D_{\bar{x}} = \pm 1$ assumes an antenna radiation angle of 90° or more, which is not given using directive radiators, e.g. horn antennas. In that case, the range of plausible derivative values can further be restricted. Assuming a maximum antenna radiation angle α and a distance between transmitter and receiving antenna of $2d$ the minimum reasonable value $D_{\min}$ can easily be defined by

$$D_{\min} = \frac{d}{\sin \alpha} \tag{6}$$

Wave fronts with lower D values would imply specular points which are located outside the antenna beam and, therefore, can be ignored [98].

Furthermore, a maximum distance derivative $D_{\bar{x}}$ depending on α, d and D can be established:

$$D_{\bar{x}\max} = \frac{\sqrt{\left(L + \cos \beta \cdot \Delta x\right)^2 - \left(L + \cos \beta \cdot \Delta x\right)\cos \beta \cdot 2d + d^2} - \sqrt{\left(L - \cos \beta \cdot \Delta x\right)^2 - \left(L - \cos \beta \cdot \Delta x\right)\cos \beta \cdot 2d + d^2}}{2 \cdot \Delta x} \tag{7}$$

with the perpendicular from the reflection plane to the distant antenna $L = \cos \gamma \cdot \dfrac{\left(D^2 - d^2\right)}{D - d \sin \alpha}$, its perpendicular angle β and the reflection angle γ as depicted in Fig. 45. This value yields

$D_{\bar{x}\,\text{max}} = \sin\alpha$ for mono-static arrangements $(d = 0)$ and approaches to this value in the case of $L \gg d$, respectively. For further details of the derivation of these thresholds and reconstruction examples illustrating the accuracy enhancement due to the application of these thresholds, we refer to [98].

5.3.3. Reconstruction results

For repeatable measurements, we applied a female dressmaker torso which is filled with tissue-equivalent phantom material (Fig. 46). Based on linear and rotational scanners which can move or rotate the object and/or the antennas, several non-planar scan schemes can be realized in order to scan this torso efficiently. In the following, the results of breast shape identification based on a toroidal scan will be shown. The M-sequence radar device used has a bandwidth of 12 GHz [97].

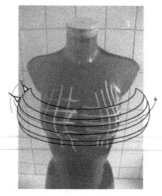

Figure 46. Female torso filled with human tissue mimicking phantom material and delineation of the toroidal scan scheme to reconstruct the chest surface

Numerical problems may arise in the calculation of derivatives from discrete data (discrete time intervals; discrete antenna positions in the space) which have to be considered for setting measurement and processing parameters. The resolutions of spatial scanning and radar signal sampling have to be harmonized carefully with each other in order to avoid derivative artifacts. The maximum possible error of the derivative is $\hat{e}(D_{\bar{x}}) = \dfrac{\Delta t \cdot v_0}{2\Delta x}$ where Δt is the time resolution of the wave front detection, Δx is the antenna displacement applied for the calculation of $D_{\bar{x}}$ and v_0 is the propagation velocity of the electromagnetic wave. Hence, it will be obvious to meet the requirement of for example $\hat{e}(D_{\bar{x}}) \leq 0.05$ (0.05 is more than 5 percent relative error with respect to $D_{\bar{x}\,\text{max}}$ for antenna beam widths < 90°!) with an antenna displacement such as $\Delta x = 2.5$ cm in air $(v_0 = c_0)$ the wave front detection has to be realized with a time accuracy of 8.33 ps which has to be provided by interpolation within the wave front detection algorithm. Higher performance requirements presuppose an even more precise wave front identification. Naturally, this is only possible if the radar device fulfills such high time stability requirements.

Figure 47 shows the UWB reconstruction results of the mentioned torso in comparison to a laser reference measurement. In order to quantify the accuracy, the distances between each calculated UWB surface point and the laser-based detected surface is calculated. The resulting mean aberration lower than 1.4 mm underlines the potential of this method. Nevertheless, it is obvious that a further enhancement of the wave front detection represents a residual challenge in order to fill in increasingly the areas of sparsely distributed surface points.

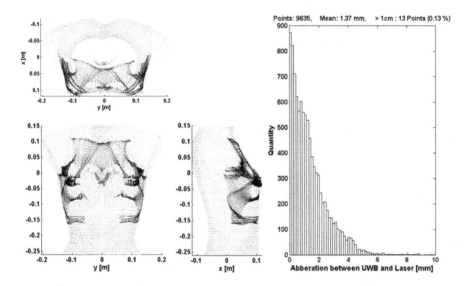

Figure 47. Exact UWB chest surface reconstruction (black) and appraisal of performance values by means of a laser reference measurement (gray) showing a mean aberration lower than 1.4 mm.

Furthermore, the applicability of a 3D-IBBST-based UWB surface reconstruction method for medical applications other than breast imaging as well as for security scenarios (under dress weapon detection) is demonstrated in [98].

5.4. Contact based breast imaging

5.4.1. UWB breast imaging in time domain

The main parts of UWB time domain imaging are the removal of clutter (also referred to as early time artifact removal) and beam-forming (also referred to as migration or back projection). Because the tumor reflections are overlapped by antenna cross-talk and skin reflection, clutter removal is a very important and critical component of signal preprocessing before beam-forming can be carried out. Most clutter removal approaches assume that the clutter appears very similar in each channel and, thus, its estimation improves with increasing channel number. It must be noted that this holds only for channels with comparable clutter parameters. That means clutter estimation and removal has to be done separately for groups consisting of only associated signals (channels with identical antenna distances and boresight angles Tx-Rx), which accomplishes this task. In scientific work on simulation, this circumstance is commonly ignored. For practical applications, however, it has to be taken into consideration.

The simplest approach is to estimate the clutter by means of the average value. Tumor reflections are assumed to appear uncorrelated in the channels and to be negligible in the averaged signal. Even though publications about advanced clutter removal algorithms emphasize the weak points of this self-evident approach, it must be noted that it works relatively robustly in the case of covering tumor response by clutter when some of the proposed alternatives are not applicable.

Image formation algorithms using time domain beam-forming can be included in the following generalized formula:

$$I(\mathbf{r}_0) = \sum_{\tau_h=-T_h/2}^{T_h/2} h(\tau_h,\mathbf{r}_0) \cdot \left(\sum_{n=1}^{N} \sum_{\tau_W=-T_w/2}^{T_w/2} w_n(\tau_w,\mathbf{r}_0) \cdot S_n(t + \tau_n(\mathbf{r}_0) + \tau_w + \tau_h) \right)^2 \tag{8}$$

where N is the number of channels, $S_n(t)$ is the clutter subtracted signal of channel n, $\mathbf{r}_0$ symbolizes the coordinates of the focal point (image position vector), $\tau_n(\mathbf{r}_0)$ is the time delay of channel n related to the focal point at $\mathbf{r}_0$ and $I(\mathbf{r}_0)$ is the back scattered energy which has to be mapped over the region of interest inside the breast. Based on two FIR filters, the different extensions of the common delay-and-sum beam former can be expressed. Path-dependent dispersion and attenuation [99], [100] can be equalized by means of $w_n(\tau_w,\mathbf{r}_0)$ which – in the simplest case - can be only a weight coefficient. Other improvements can also be included by convolution in the time domain, e.g. the cross-correlated back projection algorithm [101]. $h(\tau_h,\mathbf{r}_0)$ represents a smoothing window at the energy level or a scalar weight coefficient [102].

5.4.2. Measurement setup based on small antennas

The efficient penetration of the electromagnetic waves into the tissue and the spatial high-resolution registration of the reflected signals are crucial tasks of the antenna array design. In this regard, efficiency is not only a matter of radiation efficiency or antenna return loss, respectively. An efficient antenna array design concerning biomedical UWB imaging purposes comprises also the shape and duration of signal impulses, angle dependence of the impulse characteristics (fidelity), and the physical dimensions of the antenna. These interacting parameters are hardly to accommodate to each other within one antenna design. Generally, compromise solutions have to be found considering basic conditions of scanning, tissue properties and image processing. Here, we pursue the objective of very small antenna dimensions, short impulses and an application in direct or quasi direct contact mode. Therefore, we investigated the usability of small interfacial dipoles.

Figure 48. Small bow-ties on Rogers substrate

Initially, we used short bow-ties (Fig. 48) with the dimensions of 8 mm x 3 mm implemented on Rogers® 4003 substrate (0.5 mm) using PCB technology. Dipoles have to be fed differentially. The balanced feeding is realized by differential amplifier circuits [103].

These antennas cannot be matched over a large bandwidth, which leads to unwanted reflections between antenna and amplifier. There are two options concerning the handling of this problem: realization of a sufficient line length between antenna and amplifier (in order to gate out the reflections) or implementation of the amplifier circuits directly at the antenna feed point. On an interim basis, we pursued the first strategy using long cables between antenna and amplifier. Assuming a maximum mean tissue permittivity $\varepsilon' \leq 50$, a 70 cm cable will ensure that any reflections from inside of the breast (diameter ~ 10 cm) and unwanted reflections at the amplifier do not overlap.

As mentioned above, the contact between antennas and breast skin represents a crucial aspect for sufficient signal quality. Regarding clinical requirements (e.g. disinfection) we plan to place the antennas behind a thin examination mold. But this additional interface reduces the signal quality significantly. Therefore, a thin (~2 mm) matching layer consisting of tissue mimicking phantom material was inserted between the examination mold and the antennas in order to increase the signal energy penetrating the tissue and reduce the backward radiation (Fig. 49). The benefit achieved when using a thin contact layer was also investigated and verified by simulations (Fig. 5 in section 2.3.2).

We built up two preliminary array set-ups for phantom measurements, both including eight antennas and distributing them around a circular segment (diameter 9.5 cm) in steps of 22.5°. An array with a horizontal antenna arrangement is shown in Fig. 50. Exemplary phantom measurement results achieved with these prototypes are published in [104] and [105] and will be summarized in section 5.4.3.

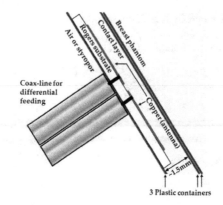

Figure 49. Schematic illustration of the contact layer filled with phantom material and mounted antennas inside

Figure 50. Antenna array: Assembly stage before casting the contact layer. The connected and affixed differential fed antennas and the container for the outer boundary of the contact layer are still visible (left panel). Finished antenna array with inserted rotatable breast phantom (right panel)

After this preliminary development stage, the differential feeding amplifier was relocated into the antenna feed point. By this step, reflections due to antenna mismatch will be avoided, and the quantity of feeding cables will be bisected, because each active antenna element can be fed single-ended (Fig. 51).

In conjunction with this enhancement, the mechanical part of the antenna array was improved. A developed slide-in mounting system (Fig. 51) allows flexible antenna application and replacement and, therefore, facilitates investigations of various Rx-Tx-arrangements without destruction and rebuild of the whole array as it is the case with the preliminary set-up shown in Fig. 50 [106].

Because the contact layer will not be hermetically sealed in this case, the chemical instable oil-gelatin phantom material cannot be used anymore for this task. Thus, investigations of alternative materials have to be considered. We propose polymer-powder composites where dielectric powders (e.g. carbon meal or barium titanate powder) will be admixed to silicone rubber. This special challenging topic is currently under investigation.

 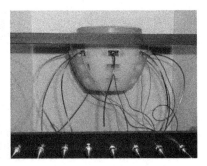

Figure 51. Photographs of an active antenna element (Rx) with 8 mm dipole with amplifier circuit board (left panel) and the slide-on mounting system for phantom measurements as well as *in vivo* measurements

5.4.3. Imaging results of phantom trials

The breast phantoms are tissue mimicking oil-gelatin phantoms according to [24] and described in section 5.2, where the dielectric properties can be adjusted by means of the oil content. For our measurements we used two types of material: 40% oil (57% water) content material mimics healthy tissue which approximately corresponds to group II of adipose-defined tissue (31%-84% adipose tissue) [106]. The 10% oil (85.5% water) content material simulates tumor tissue. Fig. 52 illustrates permittivity, attenuation losses and reflection coefficient between both tissues. In order to realize an optimal contact to the antenna array, the phantom material is filled in identical plastic containers (diameter 9.5cm) as used for the examination mold. The containers are hermetically sealed and stored in the fridge to avoid chemical instability of the phantom material. The phantoms have to be acclimatized at least 3 hours before starting the measurements.

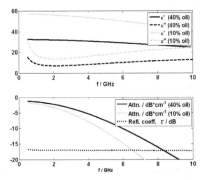

Figure 52. Dielectric values of the tissue mimicking phantom material: Permittivity (above), transmission losses per cm and reflection coefficient between them (below)

Figure 53 shows two measured signals of the proposed antennas which illustrate the appropriate time domain characteristics. The measurement through 6 cm tissue (mimicked by means of phantom material) as well as the cross-talk signal between two antennas show relatively short impulse shapes with low ringing, which is essential for UWB imaging. Including the dispersive tissue impact the spectral bulk ranges between 1 GHz and 3 GHz with a bandwidth greater than 2 GHz for both received impulses. Obviously, because of the dielectric scaling due to the direct contact between tissue and antenna, such small antennas are capable of radiating waves in a frequency range with acceptable attenuation and penetration depth.

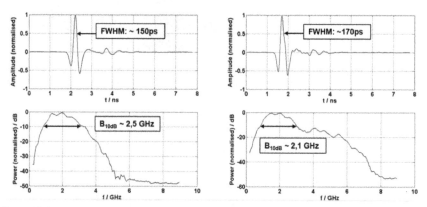

Figure 53. Measurement signals based on the described bow-ties: measurement through 6 cm tissue mimicking phantom material with 40% oil content (left panel) and cross-talk signal between adjacent antennas, separated by 2.5 cm (right panel)

During the phantom measurements, four antennas acted as receivers and are permanently connected with Rx1...Rx4 of the radar device. The transmitter signal was connected to one of 4 transmitter antennas by a coaxial switch matrix. Thus, 16 signal channels could be achieved without rearrangement. Their angles between the boresight directions of Tx and Rx differed in the range 22.5 - 157.5°. Because this amount of signal channels is insufficient for high-resolution imaging, we had to consider robust and reproducible mechanical scanning to achieve a sufficient number of channels. In order to simulate antenna rotation, the phantoms were rotated in steps of 11.25°. This resulted in 512 signals (16 channels x 32 rotations) which could be included into the imaging process of one phantom.

Figure 54 shows exemplary imaging results of the described breast phantoms applying the presented measuring set-up and time domain beam-forming. Despite the relatively low dielectric contrast between both tissue simulations, the tumor inclusions can clearly be identified. The highest interferences (side lobes) are about 11dB (15mm tumor) and around 7dB (10mm tumor) lower than the tumor representation. Additionally, the lower panels of Fig. 54 illustrate the capability of localization and differentiation between multiple tumors, for example two 15 mm tumors with a distance of 30 mm between them. Despite of the relatively low dielectric contrast between healthy and cancerous tissue mimicking phantom material, the tumors can be detected and separated.

The results underline that small dipoles can be profitably applied for UWB breast imaging. The impressive identification of the tumor surrogates promises also the detection of weaker dielectric contrasts. On the other hand, it must be noted that the tumor surrounding tissue imitation is completely homogeneous which does not correspond to reality. Therefore, our breast phantoms must be enhanced in the future toward a better approximation of the breast tissue heterogeneity.

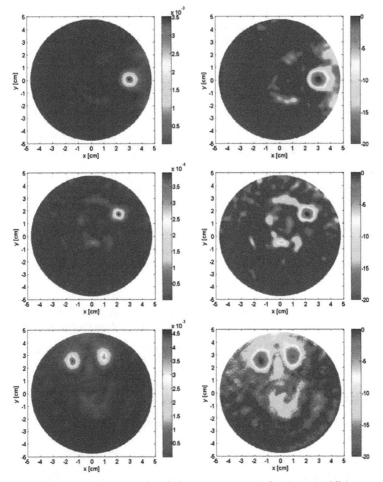

Figure 54. UWB images of phantom trials including a 15mm (top) and a 10 mm (middle) tumor surrogate as well as two 15 mm tumor surrogates, separated by 30 mm (below). Left: linear energy scale; Right: logarithmic scale in dB.

6. Conclusions

In this chapter, we dealt with ultra wideband sensing in medical engineering, i.e. using electromagnetic waves of large bandwidth for probing the human body and biological tissue. Sufficient penetration of the human body combined with antennas of manageable size were our major concern. Also, the frequency band from 1 GHz to 5...8 GHz turned out to be best suited for our purposes. By using active or dielectrically scaled antennas for this frequency range, they can be built sufficiently small. Wave propagation at these frequencies is mostly influenced by water, the most abundant component of biological tissue. The effect of salt becomes less detrimental above 1 GHz. Above 5...8 GHz, however, water absorption will drastically increase the propagation losses. The given frequency band also provides acceptable resolution for microwave imaging and ample micro-Doppler sensitivity.

For experimental investigations, we exploited ultra-wideband pseudo-noise devices. They provide probing signals of very low power, thus avoiding damages to biological tissue. Furthermore, they provide sufficient dynamic range, measurement speed and short term stability for super resolution techniques of microwave imaging and weak-motion tracking.

We demonstrated medical applications of ultra-wideband sensing by three distinctive examples, each standing for a specific class of applications.

1. Contact-based measurements (impedance spectroscopy) aimed to estimate tissue permittivity. This mainly gives some hints on water concentration and water bonds. At lower frequencies, other molecules will also leave their traces in the measured impedance value.

2. Remote motion tracking of organs inside the human body like cardio-pulmonary activity for example of motion correction for magnetic resonance imaging. Remote vital sign detection is a related topic with relaxed conditions referred to tracking precision but increased requirements with respect to area coverage. The analysis of cardiac mechanics for separate heart region accessible by stand-alone UWB radar or in conjunction with the electrical activity from the ECG contains valuable diagnostic information, e.g. for infarction detection, as ischemic tissue shows a modified contraction pattern.

Remote or contact-based microwave imaging of inner organs or malignant tissue, for example the detection of breast tumors.

Author details

Ingrid Hilger, Katja Dahlke, Gabriella Rimkus and Christiane Geyer
Jena University Hospital, Germany

Frank Seifert, Olaf Kosch, Florian Thiel
Physikalisch-Technische Bundesanstalt Berlin, Germany

Matthias Hein, Francesco Scotto di Clemente, Ulrich Schwarz, Marko Helbig, Jürgen Sachs
Ilmenau University of Technology, Germany

Acknowledgement

This work was supported by the German Science Foundation (DFG) in the framework of the priority program UKoLoS (SPP 1202), project acronym ultraMEDIS. The authors appreciate the valuable contributions made by R. Herrmann, P. Rauschenbach and K. Schilling for sensor development, and helpful discussions; K. Borkowski and E. Hamatschek for electronic and mechanical component manufacturing; Ralf Stephan for his support in antenna design and measurement; Hartmut Günther and Stefan Barth for the manufacture of the ceramic antennas; Marina Sieler and Uwe Genatis for the galvanic metallization of the ceramic and MRI compatible antennas.

7. References

[1] U. Pliquett, "Electricity and biology," in 11th International Biennial Baltic Electronics Conference, BEC 2008, pp. 11-20.

[2] K. Nowak, W. Gross, K. Nicksch et al., "Intraoperative lung edema monitoring by microwave reflectometry," Interactive Cardiovascular and Thoracic Surgery, pp. 540-544, 2011.

[3] J. Sachs, E. Zaikov, M. Helbig et al., "Trapped Victim Detection by Pseudo-Noise Radar," in 2011 International Conference on Wireless Technologies for Humanitarian Relief (ACWR 2011) Amritapuri, Kollam, Kerala, India, 2011, pp. 265-272.

[4] E. Zaikov, and J. Sachs, "UWB radar for detection and localization of trapped people," Ultra Wideband, B. Lembrikov, ed., Rijeka, Croatia: Scivo, 2010.

[5] J. Sachs, M. Helbig, R. Herrmann et al., "Merkmalsextraktion und semantische Integration von Ultrabreitband-Sensoren zur Erkennung von Notfällen," in 3. Deutscher AAL-Kongress 2010 Assistenzsysteme im Dienste des Menschen, Berlin, 2010.

[6] R. Herrmann, J. Sachs, and F. Bonitz, "On benefits and challenges of person localization using ultra-wideband sensors," in Indoor Positioning and Indoor Navigation (IPIN), 2010 International Conference on, 2010, pp. 1-7.

[7] R. Herrmann, J. Sachs, K. Schilling et al., "12-GHz Bandwidth M-Sequence Radar for Crack Detection and High Resolution Imaging," in International Conference on Ground Penetrating Radar (GPR), Birmingham, UK, 2008.

[8] R. Herrmann, "M-sequence based ultra-wideband radar and its application to crack detection in salt mines" PhD-Thesis, Faculty of Electrical Engineering and Information Technology, Ilmenau University of Technology (Germany), Ilmenau, 2011.

[9] J. Sachs, "Ultra-Wideband Short-Range Sensing - Theory, Sensors, Applications", Berlin: Wiley-VCH, 2012.

[10] M. A. Hein, C. Geyer, M. Helbig et al., "Antennas for ultra-wideband medical sensor systems," in 3rd European Conference on Antennas and Propagation, EuCAP 2009, 2009, pp. 1868-1872.

[11] R.C. Hansen, "Fundamental Limitations in Antennas", in Proc. of the IEEE, vol. 69, n. 2, pp. 170-182, Feb. 1981

[12] U. Schwarz, M. Helbig, J. Sachs, R. Stephan, M.A. Hein, "Design and application of dielectrically scaled double-ridged horn antennas for biomedical UWB radar applications", 2009 IEEE International Conference on Ultra-Wideband ICUWB 2009, pp. 150-154

[13] F. Thiel, M.A. Hein, J. Sachs, U. Schwarz and F. Seifert: "Physiological signatures monitored by ultra-wideband-radar validated by magnetic resonance imaging", IEEE International Conference on Ultra-Wideband, ICUWB 2008, pp. 105-108

[14] The Visible Human Project, U.S. National Library of Medicine, 8600 Rockville Pike, Bethesda, MD 20894

[15] U. Schwarz, M. Helbig, J. Sachs, F. Seifert, R. Stephan, F. Thiel, M.A. Hein, "Physically small and adjustable double-ridged horn antenna for biomedical UWB radar applications", IEEE International Conference on Ultra-Wideband, ICUWB 2008, pp. 5-8

[16] H. Schantz, "Ultrawideband Antennas", Artech House, Inc., 2005

[17] MRA Laboratories, Adams, MA, USA. Available online: www.mralabs.com (accessed May 2012)

[18] J. Krupka, R.G. Geyer, J. Baker-Jarvis, J. Ceremuga, "Measurements of the complex permittivity of microwave circuit board substrates using split dielectric resonator and reentrant cavity techniques", 7th International Conference on Dielectric Materials, Measurements and Applications, (Conf. Publ. No. 430), 1996, pp. 21-24

[19] R.E. Collin, "Foundations for microwave engineering", 2nd ed., McGraw-Hill, 1992

[20] J. Krupka, "Computations of Frequencies and Intrinsic Q Factors of TE0nm Modes of Dielectric Resonators", IEEE Transactions on Microwave Theory and Techniques, 1985, pp. 247-277

[21] F. Scotto di Clemente, R. Stephan, and M. Hein, "Compact permittivity-matched ultra-wideband antennas for biomedical imaging", Topical Conference on Antennas and Propagation in Wireless Communications (APWC), 2011 IEEE-APS, pp. 858-860

[22] F. Scotto di Clemente, M. Helbig, J. Sachs, U. Schwarz, R. Stephan, M.A. Hein, "Permittivity-matched compact ceramic ultra-wideband horn antennas for biomedical diagnostics", Proceedings of the 5th European Conference on Antennas and Propagation (EUCAP), pp. 2386-2390

[23] U. Schwarz, R. Stephan, M.A. Hein, "Miniature double-ridged horn antennas composed of solid high-permittivity sintered ceramics for biomedical ultra-wideband radar applications", 2010 IEEE Antennas and Propagation Society International Symposium (APSURSI), 2010, pp. 1-4

[24] M. Lazebnik, E. Madsen EL, G. Frank, S. Hagness, "Tissue-mimicking phantom materials for narrowband and ultra-wideband microwave applications", Physics in Medicine and Biology 2005; 50:4245-4258.

[25] C. Geyer, "Wechselwirkungen von elektromagnetischen Wellen im Radiofrequenzbereich mit biologischen Systemen und lebenden Organismen," Dissertation, Biologisch-Pharmazeutische Fakultät, Friedrich-Schiller-Universität Jena, 2011.

[26] U. Schwarz, R. Stephan, M.A. Hein, "Experimental validation of high-permittivity ceramic double-ridged horn antennas for biomedical ultra-wideband diagnostics", IEEE

International Conference on Wireless Information Technology and Systems (ICWITS), 2010, pp. 1-4.

[27] M. Hein, M. Helbig, M. Kmec, J. Sachs, F. Scotto di Clemente, R. Stephan, M. Hamouda, T. Ussmueller, R. Weigel, M. Robens, R. Wunderlich, S. Heinen, "Ultra-Wideband Active Array Imaging for Biomedical Diagnostics", Topical Conference on Antennas and Propagation in Wireless Communications (APWC), 2012 IEEE-APS (Accepted).

[28] F. Scotto di Clemente, R. Stephan, U. Schwarz, M.A. Hein, " Miniature body-matched double-ridged horn antennas for biomedical UWB imaging", Topical Conference on Antennas and Propagation in Wireless Communications (APWC), 2012 IEEE-APS (Accepted)

[29] F. Thiel, M. Hein, J. Sachs, U. Schwarz, and F. Seifert, "Combining magnetic resonance imaging and ultrawideband radar: A new concept for multimodal biomedical imaging," Rev. Sci. Instrum., vol. 80, no. 1, 2009.

[30] O. Kosch, F. Thiel, F. Scotto di Clemente, M.A. Hein, F. Seifert, "Monitoring of human cardio-pulmonary activity by multi-channel UWB-radar", Topical Conference on Antennas and Propagation in Wireless Communications (APWC), 2011 IEEE-APS, pp.382-385

[31] U. Schwarz, F. Thiel, F. Seifert, R. Stephan, M. Hein, "Magnetic resonance imaging compatible ultra-wideband antennas," in 3rd Eur. Conf. on Antennas and Propagation, 2009, pp. 1102–1105.

[32] U. Schwarz, F. Thiel, F. Seifert, R. Stephan, M.A. Hein, "Ultra-Wideband Antennas for Magnetic Resonance Imaging Navigator Techniques", IEEE Trans. on Antennas and Propag, vol. 58, no. 6, June 2010.

[33] D. Miklavcic, N. Pavselj, and F. X. Hart, "Electrical properties of tissues," *Wiley Encyclopedia of Biomedical Engineering*, M. Akay, ed.: Wiley-Interscience, 2006.

[34] C. Gabriel, "Dielectric properties of biological tissue: Variation with age," *Bioelectromagnetics*, pp. S12-S18, 2005.

[35] G. H. Markx, and C. L. Davey, "The dielectric properties of biological cells at radiofrequencies: Applications in biotechnology," *Enzyme and Microbial Technology*, vol. 25, no. 3-5, pp. 161-171, Aug, 1999.

[36] R. Pethig, "Dielectric-Properties of Biological-Materials - Biophysical and Medical Applications," *IEEE Transactions on Electrical Insulation,* vol. 19, no. 5, pp. 453-474, 1984.

[37] H. P. Schwan, "Analysis of Dielectric Data - Experience Gained with Biological-Materials," *IEEE Transactions on Electrical Insulation,* vol. 20, no. 6, pp. 913-922, Dec, 1985.

[38] H. P. Schwan, and C. F. Kay, "The Conductivity of Living Tissues," *Annals of the New York Academy of Sciences,* vol. 65, no. 6, pp. 1007-1013, 1957.

[39] A. S. Arnoux, L. Preziosi-Belloy, G. Esteban *et al.,* "Lactic acid bacteria biomass monitoring in highly conductive media by permittivity measurements," *Biotechnology Letters,* vol. 27, no. 20, pp. 1551-1557, Oct, 2005.

[40] J. Yang, Y. Huang, X. J. Wang *et al.,* "Dielectric properties of human leukocyte subpopulations determined by electrorotation as a cell separation criterion," *Biophysical Journal,* vol. 76, no. 6, pp. 3307-3314, Jun, 1999.

[41] H. P. Schwan, "Mechanisms responsible for electrical properties of tissues and cell suspensions," *Med Prog Technol*, vol. 19, no. 4, pp. 163-5, 1993.

[42] S. R. Smith, and K. R. Foster, "Dielectric properties of low-water-content tissues," *Phys Med Biol*, vol. 30, no. 9, pp. 965-73, Sep, 1985.

[43] K. R. Foster, and H. P. Schwan, "Dielectric-Properties of Tissues and Biological-Materials - a Critical-Review," *Critical Reviews in Biomedical Engineering*, vol. 17, no. 1, pp. 25-104, 1989.

[44] J. A. Evans, S. L. Russell, C. James *et al.*, "Microbial contamination of food refrigeration equipment," *Journal of Food Engineering*, vol. 62, no. 3, pp. 225-232, May, 2004.

[45] M. Grossi, A. Pompei, M. Lanzoni *et al.*, "Total Bacterial Count in Soft-Frozen Dairy Products by Impedance Biosensor System," *IEEE Sensors Journal*, vol. 9, no. 10, pp. 1270-1276, Oct, 2009.

[46] Y. J. Chang, A. D. Peacock, P. E. Long *et al.*, "Diversity and characterization of sulfate-reducing bacteria in groundwater at a uranium mill tailings site," *Applied and Environmental Microbiology*, vol. 67, no. 7, pp. 3149-3160, Jul, 2001.

[47] S. K. Haack, L. R. Fogarty, T. G. West *et al.*, "Spatial and temporal changes in microbial community structure associated with recharge-influenced chemical gradients in a contaminated aquifer," *Environmental Microbiology*, vol. 6, no. 5, pp. 438-448, May, 2004.

[48] G. Haferburg, D. Merten, G. Buchel *et al.*, "Biosorption of metal and salt tolerant microbial isolates from a former uranium mining area. Their impact on changes in rare earth element patterns in acid mine drainage," *Journal of Basic Microbiology*, vol. 47, no. 6, pp. 474-484, Dec, 2007.

[49] S. E. Beekmann, D. J. Diekema, K. C. Chapin *et al.*, "Effects of rapid detection of bloodstream infections on length of hospitalization and hospital charges," *Journal of Clinical Microbiology*, vol. 41, no. 7, pp. 3119-3125, Jul, 2003.

[50] K. Mishima, A. Mimura, Y. Takahara *et al.*, "Online Monitoring of Cell Concentrations by Dielectric Measurements," *Journal of Fermentation and Bioengineering*, vol. 72, no. 4, pp. 291-295, 1991.

[51] C. M. Harris, R. W. Todd, S. J. Bungard *et al.*, "Dielectric Permittivity of Microbial Suspensions at Radio Frequencies - a Novel Method for the Real-Time Estimation of Microbial Biomass," *Enzyme and Microbial Technology*, vol. 9, no. 3, pp. 181-186, Mar, 1987.

[52] M. Jonsson, K. Welch, S. Hamp *et al.*, "Bacteria counting with impedance spectroscopy in a micro probe station," *Journal of Physical Chemistry B*, vol. 110, no. 20, pp. 10165-10169, May 25, 2006.

[53] E. Benoit, A. Guellil, J. C. Block *et al.*, "Dielectric permittivity measurement of hydrophilic and hydrophobic bacterial suspensions: a comparison with the octane adhesion test," *Journal of Microbiological Methods*, vol. 32, no. 3, pp. 205-211, May, 1998.

[54] Dahlke, K., Geyer, C.; Dees, S.; Helbig, M.; Sachs, J.; Scotto di Clemente, F.; Hein, M.; Kaiser, W.A.; Hilger, I.,"Effects of cell structure of Gram-positive and Gramnegative bacteria based on their dielectric properties," *The 7th German Microwave Conference (GeMiC), 2012*, 12-14 March 2012.

[55] F. Jaspard, and M. Nadi, "Dielectric properties of blood: an investigation of temperature dependence," *Physiological Measurement*, vol. 23, no. 3, pp. 547-554, Aug, 2002.

[56] H. F. Cook, "A Comparison of the Dielectric Behaviour of Pure Water and Human Blood at Microwave Frequencies," *British Journal of Applied Physics*, vol. 3, no. Aug, pp. 249-255, 1952.

[57] H. P. Schwan, and K. Li, "Capacity and Conductivity of Body Tissues at Ultrahigh Frequencies," *Proceedings of the Institute of Radio Engineers*, vol. 41, no. 12, pp. 1735-1740, 1953.

[58] H. P. Schwan, and K. R. Foster, "Rf-Field Interactions with Biological-Systems - Electrical-Properties and Biophysical Mechanisms," *Proceedings of the IEEE*, vol. 68, no. 1, pp. 104-113, 1980.

[59] L. Schmuntzsch, "Thermische Abhängigkeit der Permittivität in biologischem Gewebe," Diploma, Fachbereich Medizintechnik/ Biotechnologie, Fachhochschule Jena, Jena, 2010.

[60] A. Mashal, B. Sitharaman, X. Li *et al.*, "Toward Carbon-Nanotube-Based Theranostic Agents for Microwave Detection and Treatment of Breast Cancer: Enhanced Dielectric and Heating Response of Tissue-Mimicking Materials," *IEEE Transactions on Biomedical Engineering*, vol. 57, no. 8, pp. 1831-1834, Aug, 2010.

[61] Noeske R., Seifert F., Rhein K.H., Rinneberg H., "Human cardiac imaging at 3T using phased array coils", *Magn. Reson. Med.* 44, 978-82 (2000)

[62] Haacke E.M., Li D., Kaushikkar S., "Cardiac MR Imaging: Principles and Techniques". *Top Magn Reson Imaging* 7:200-17, (1995).

[63] Dieringer M.A., Renz W., Lindel T., Seifert F., Frauenrath T., von Knobelsdorff-Brenkenhoff F., Waiczies H., Hoffmann W., Rieger J., Pfeiffer H., Ittermann B., Schulz-Menger J., Niendorf T., "Design and application of a four-channel transmit/receive surface coil for functional cardiac imaging at 7T", *Journal of Magnetic Resonance Imaging* 33 (3): 736-741 (2011-03).

[64] Nijm G.M., Swiryn S., Larson A.C., and Sahakian A.V., "Characterization of the Magnetohydrodynamic Effect as a Signal from the Surface Electrocardiogram during Cardiac Magnetic Resonance Imaging", in *Computers in Cardiology*, Valencia, Spain, p. 269–272, (2006)

[65] Frauenrath T., Hezel F., Renz W., de Geyer T., Dieringer M., von Knobelsdorff-Brenkenhoff F., Prothmann M., Schulz-Menger J., Niendorf T., "Acoustic cardiac triggering: a practical solution for synchronization and gating of cardiovascular magnetic resonance at 7 Tesla", *Journal of Cardiovascular Magnetic Resonance* 12 (1): 67 (2010-11-16).

[66] van Geuns R.J., PA. Wielopolski, HG. de Bruin, et al: "MR coronary angiography with breath-hold targeted volumes: preliminary clinical results", *Radiology* (2000); 217:270–7

[67] Hinks, R.S.. "Monitored echo gating for reduction of motion artifacts". *Magn Reson Imaging*, 6(48), (1988).

[68] Thiel F., Kreiseler D., Seifert F., "Non-contact detection of myocardium's mechanical activity by ultra-wideband RF-radar and interpretation applying electrocardiography", *Rev. Sci. Instrum.*, vol. 80, 11, 114302, ISSN 0034-6748, Melville, NY: American Institute of Physics (AIP), 12 pages, (2009)

[69] Morguet A.J.; Behrens S., Kosch O., Lange C., Zabel M., Selbig D., Munz D.L., Schultheiss H.P., Koch H.: "Myocardial viability evaluation using magnetocardiography in patients with coronary artery disease"; *Coron Artery Dis* 15, 2004: 155 – 162

[70] Muehlsteff J., Thijs J., Pinter R., Morren G., Muesch G., "A handheld device for simultaneous detection of electrical and mechanical cardio-vascular activities with synchronized ECG and CW-Doppler Radar", *IEEE Int. Conf. Engineering in Medicine and Biology Society* (EMBS) 22-26 Aug. (2007), 5758 – 5761

[71] Thiel F. and Seifert F., "Non-invasive probing of the human body with electromagnetic pulses: Modelling of the signal path", *J. Appl. Phys.*, vol.105, issue 4, 044904, ISSN 0021-8979, Melville, NY, American Institute of Physics (AIP), 8 pages, (2009)

[72] Thiel F. and Seifert F., "Physiological signatures reconstructed from a dynamic human model exposed to ultra-wideband microwave signals", *Frequenz, Journal of RF/Microwave-Engineering, Photonics and Communications*, vol. 64, no. 3-4, 34-41, ISSN 0016-1136, (2010).

[73] Thiel F., Kosch O., Seifert F., "Ultra-wideband Sensors for Improved Magnetic Resonance Imaging, Cardiovascular Monitoring and Tumour Diagnostics", Special Issue *"Sensors in Biomechanics and Biomedicine"*, 10, 12, 10778-10802, doi:10.3390/s101210778, ISSN 1424-8220, 25 pages, (2010)

[74] Kosch O., Thiel F., Yan D.D., and Seifert F., "Discrimination of respirative and cardiac displacements from ultra-wideband radar data", *Int. Biosignal Processing Conf.* (Biosignal 2010), 14th – 16th July; Berlin, Germany, (2010)

[75] Christ A. et al., "The virtual family – development of surface-based anatomical models of two adults and two children for dosimetric simulations", *Phys. Med. Biol.* 55, N23-N38, (2010)

[76] Sachs J., Peyerl P., Wöckel S., et al., "Liquid and moisture sensing by ultra-wideband pseudo-noise sequence signals", *Meas. Sci. Technol.*, vol. 18, no. 4, pp. 1074-1088, April 2007.

[77] Kosch O., Thiel F., Schneider S., Ittermann B., and Seifert F., "Verification of contactless multi-channel UWB navigator by one dimensional MRT", *Proc. Intl. Soc. Mag. Reson. Med.* 20, (ISMRM), Melbourne, Australia, p.601, (2012)

[78] Ziehe A. and Müller K.R., TDSEP - "An efficient algorithm for blind separation using time structure", *Proc. Int. Conf. Artifical Neural Networks* (ICANN'98), Skövde, Sweden, pp.675-680, (1998).

[79] Sander T.H., Lueschow A., Curio G., Trahms L., "Time delayed decorrelation for the identification of the cardiac artifact in MEG data", *Biomed. Tech.* 2002, 47(1/2), pp. 573-576, (2002).

[80] Kosch O., Schneider S., Ittermann B., and Seifert F., "Self-adjusting multichannel UWB radar for cardiac MRI", *Proc. Intl. Soc. Mag. Reson. Med.* 20, (ISMRM), Melbourne, Australia, p.2467, (2012)

[81] Kosch O., Thiel F., Ittermann B., and Seifert F., "Non-contact cardiac gating with ultra-wideband radar sensors for high field MRI", *Proc. Intl. Soc. Mag. Reson. Med.* 19, (ISMRM), Montreal, Canada, ISSN 1545-4428, p.1804, (2011)

[82] M. Helbig, "Mikrowellen- Ultrabreitband- und THz-Bildgebung," in O. Dössel, T. M. Buzug, (Hrsg.) "Medizinische Bildgebung", Lehrbuchreihe der Biomedizinischen Technik, Bd. 7, Berlin, de Gruyter, 2012

[83] E. C. Fear, S. C. Hagness, P. M. Meaney et al., "Enhancing breast tumor detection with near-field imaging," Microwave Magazine, IEEE, vol. 3, no. 1, pp. 48-56, 2002.

[84] P. M. Meaney, M. W. Fanning, D. Li et al., "A clinical prototype for active microwave imaging of the breast," IEEE Transactions on Microwave Theory and Techniques, vol. 48, no. 11, pp. 1841-1853, Nov, 2000.

[85] M. Klemm, I. J. Craddock, J. A. Leendertz et al., "Radar-Based Breast Cancer Detection Using a Hemispherical Antenna Array - Experimental Results," IEEE Transactions on Antennas and Propagation, vol. 57, no. 6, pp. 1692-1704, 2009.

[86] M. Klemm, I. J. Craddock, J. A. Leendertz et al., "Clinical trials of a UWB imaging radar for breast cancer," Proceedings of the Fourth European Conference on Antennas and Propagation (EuCAP), 2010, pp. 1-4.

[87] J. Sachs, "M-sequence radar," Ground penetrating radar, D. J. Daniels, ed., pp. 225-237, London: The Institution of Engineering and Technology, 2004.

[88] U. Kaatze, "Complex permittivity of water as a function of frequency and temperature," J Chem Eng Data, vol. 34, no. 4, pp. 371-374, Oct, 1989.

[89] T. C. Williams, J. Bourqui, T. R. Cameron et al., "Laser Surface Estimation for Microwave Breast Imaging Systems," IEEE Transactions on Biomedical Engineering, vol. 58, no. 5, pp. 1193-1199, 2011.

[90] D. W. Winters, J. D. Shea, P. Kosmas et al., "Three-Dimensional Microwave Breast Imaging: Dispersive Dielectric Properties Estimation Using Patient-Specific Basis Functions," IEEE Transactions on Medical Imaging, vol. 28, no. 7, pp. 969-981, 2009.

[91] T. Sakamoto, and T. Sato, "A target shape estimation algorithm for pulse radar systems based on boundary scattering transform," IEICE Transactions on Communications, vol. E87b, no. 5, pp. 1357-1365, May, 2004.

[92] S. Kidera, Y. Kani, T. Sakamoto et al., "Fast and accurate 3-D imaging algorithm with linear array antennas for UWB pulse radars," IEICE Transactions on Communications, vol. E91b, no. 8, pp. 2683-2691, Aug, 2008.

[93] M. Helbig, M. A. Hein, U. Schwarz et al., "Preliminary investigations of chest surface identification algorithms for breast cancer detection," IEEE International Conference on Ultra-Wideband, ICUWB 2008, pp. 195-198.

[94] M. Helbig, C. Geyer, M. Hein et al., "A Breast Surface Estimation Algorithm for UWB Microwave Imaging," IFMBE Proceedings 4th European Conference of the International Federation for Medical and Biological Engineering, J. Sloten, P. Verdonck, M. Nyssen et al., eds., pp. 760-763: Springer Berlin Heidelberg, 2009.

[95] M. Helbig, C. Geyer, M. Hein et al., "Improved Breast Surface Identification for UWB Microwave Imaging," IFMBE Proceedings World Congress on Medical Physics and Biomedical Engineering, September 7 - 12, 2009, Munich, Germany, O. Dössel and W. C. Schlegel, eds., pp. 853-856: Springer Berlin Heidelberg, 2009.

[96] S. Hantscher, B. Etzlinger, A. Reisenzahn *et al.*, "A Wave Front Extraction Algorithm for High-Resolution Pulse Based Radar Systems," *IEEE International Conference on Ultra-Wideband, ICUWB 2007*, pp. 590-595.

[97] J. Sachs, R. Herrmann, M. Kmec *et al.*, "Recent Advances and Applications of M-Sequence based Ultra-Wideband Sensors," *IEEE International Conference on Ultra-Wideband, ICUWB 2007*, pp. 50-55.

[98] M. Helbig, " UWB for Medical Microwave Breast Imaging," in J. Sachs, *"High-Resolution Short-Range Sensing - Basics, Principles and Applications of Ultra Wideband Sensors*, Berlin: Wiley-VCH, 2012

[99] E. J. Bond, X. Li, S. C. Hagness *et al.*, "Microwave imaging via space-time beamforming for early detection of breast cancer," *IEEE Transactions on Antennas and Propagation*, vol. 51, no. 8, pp. 1690-1705, Aug, 2003.

[100] Y. Xie, B. Guo, L. Xu *et al.*, "Multistatic Adaptive Microwave Imaging for Early Breast Cancer Detection," *IEEE Transactions on Biomedical Engineering*, vol. 53, no. 8, pp. 1647-1657, 2006.

[101] R. Zetik, J. Sachs, and R. Thoma, "Modified cross-correlation back projection for UWB imaging: numerical examples," *IEEE International Conference on Ultra-Wideband, ICU 2005*, pp. 650-654.

[102] M. Klemm, I. J. Craddock, J. A. Leendertz *et al.*, "Improved Delay-and-Sum Beamforming Algorithm for Breast Cancer Detection," *International Journal of Antennas and Propagation*, vol. 2008.

[103] M. Kmec, M. Helbig, J. Sachs et al., "Integrated ultra-wideband hardware for MIMO sensing using pn-sequence approach," *IEEE International Conference on Ultra-Wideband, ICUWB 2012, Syracuse, USA*, 17-20 Sept. 2012

[104] M. Helbig, I. Hilger, M. Kmec, *et al.*, "Experimental phantom trials for UWB breast cancer detection", *Proc. German Microwave Conf.*, Ilmenau, Germany, pp. 1-4, 12-14 March 2012.

[105] M. Helbig, M. Kmec, J. Sachs *et al.*, "Aspects of antenna array configuration for UWB breast imaging," Proc. European Conf. on Antennas and Propagation, EuCAP 2012, Prague, pp. 1-4, 26-30 March 2012.

[106] M. Helbig, M. A. Hein, R. Herrmann *et al.*, "Experimental active antenna measurement setup for UWB breast cancer detection," *IEEE International Conference on Ultra-Wideband, ICUWB 2012, Syracuse, USA*, 17-20 Sept. 2012

[107] M. Lazebnik, D. Popovic, L. McCartney *et al.*, "A large-scale study of the ultrawideband microwave dielectric properties of normal, benign and malignant breast tissues obtained from cancer surgeries," *Physics in Medicine and Biology*, vol. 52, no. 20, pp. 6093-6115, Oct 21, 2007.

Chapter 4

HaLoS – Integrated RF-Hardware Components for Ultra-Wideband Localization and Sensing

Stefan Heinen, Ralf Wunderlich, Markus Robens, Jürgen Sachs, Martin Kmec, Robert Weigel, Thomas Ußmüller, Benjamin Sewiolo, Mohamed Hamouda, Rolf Kraemer, Johann-Christoph Scheytt and Yevgen Borokhovych

Additional information is available at the end of the chapter

1. Introduction

Ultra-Wideband (UWB) sensors exploit very weak electromagnetic waves within the lower microwave range for sounding the objects or processes of interest. The interaction of electromagnetic waves with matter provides interesting options to gain information from a great deal of different scenarios. To mention only a few, it enables the assessment of the state of building materials and constructions, the investigation of biological tissue, the detection and localization of persons buried by rubble after an earthquake or unauthorized people hidden behind walls, and much more [1]. The advantage of such methods consists in their non-destructive and continuously running measurement procedure which may work at high speed and in contactless fashion.

Sensors applying electromagnetic interactions with the test object have been in use for a long time. However, most of such sensors are restricted to a relatively narrow bandwidth and, consequently, they can provide only a small amount of information about the test object. Sophisticated data processing supposed, UWB sensors may be able to provide more information and, therefore, to reduce ambiguities which are inherently part of indirect measurement methods such as electromagnetic sensing.

Depending on the actual tasks, the requirements on the sensing system may be quite different, such as the optimum operational frequency band, measurement speed, sensitivity, system costs, reliability, power consumption etc. There are several UWB sensing principles known, each having specific advantages and disadvantages. Generally, one can state that the usability of UWB-sensors will be largely improved with increasing degree of system integration regardless of the sensor principle. The HaLoS-project addresses this topic by investigating general purpose UWB sub-modules like amplifiers, ADCs, fast processing units etc. as well as an integration-friendly sensor concept based on ultra-wideband pseudo-noise codes.

The chapter is organized as follows. First, the most important performance figures of UWB sensors are introduced. Second, we give an overview of various UWB-sensor principles recently in use and explain the UWB pseudo-noise concept. Then, we address some specific topics like wideband receiver circuits, transmitter circuits and high-speed data capture. Finally, some aspects of monolithically integrated UWB-sensors are discussed.

2. Properties and basic concepts of UWB-sensors

2.1. Key figures of UWB-sensors

The UWB sensor configuration may be determined by demands which are guided by two different and partially conflicting aspects. On the one hand, these are the UWB radiation rules, and on the other one, we have to respect the physical constraints of the sensing problem. The radiation rules, which are not unique within different regions of the world, mainly limit spectral power emission, restrict the operation frequency band and require sounding signals of large instantaneous bandwidth. Seen from a physical point of view, we need an adequate operational frequency band which provides reasonable interaction between the sounding signal and the object of interest. This may lead to conflicting situations with the radiation rule for sensing tasks requiring wave penetration like through-wall radar or medical imaging. Thus, one has to search a proper compromise in the case of frequency mask violation. Though UWB sensors are banned from long-range applications due to low-radiation power, they promote biological and medical sensing since the target exposition is harmless. Furthermore, the interaction between sounding wave and target is based on linear phenomena. Hence, the sounding bandwidth may be provided instantaneously (complying with FCC or ECC radiation rules) or sequentially (violating these radiation rules) without affecting the measurement results as long as the scenario under test behaves stationary during the measurement. This paper is focused on techniques for information capture by exploiting electromagnetic interactions. Hence, we do not exclude sensor principles or frequency bands violating UWB radiation rules from our further discussions.

Spectral band and related parameters: As frequency diversity is a key issue of unambiguous information gathering by electric sensors, the widths and the occupation density of the spectral sounding band is of major interest. For the sake of brevity, we will deal here only with baseband signals (see [2] for deeper discussions) which we characterize by their two-sided bandwidth B that can be linked to typical time domain parameters:

$$B \approx \begin{cases} t_w^{-1} & \text{for pulse shaped signal} \\ \tau_{coh}^{-1} & \text{for CW signal} \end{cases} \tag{1}$$

Here, t_w represents the width of a pulse, and τ_{coh} is the coherence time of a random or pseudo-random signal (i.e. the width of the auto-correlation function). The occupation density of the frequency band is given by the line spacing Δf which is either determined by the repetition rate f_0 of a periodic sounding signal (t_P - period duration) or via the Fourier Transform by the observation interval T of non-periodic signals:

$$\Delta f = \begin{cases} f_0 = t_P^{-1} & \text{periodic signal} \\ T^{-1} & \text{non-periodic signal} \end{cases} \qquad (2)$$

As non-periodic signals are quite unusual in UWB sensing, we will avoid discussing them. The line spacing Δf gives the frequency resolution of the sensor or it determines the maximum observable length $T_W = \Delta f^{-1}$ of the impulse response $g(t)$ of a scenario under test. If $g(t)$ does not settle down within $T_w = t_P$, we have to anticipate time aliasing.

In the case of UWB radar sensing, we can convert (1) and (2) into corresponding spatial parameters. One of them assigns the range resolution δ_r, i.e. the capability of the radar to separate two close point targets of identical reflectivity. We will refer to the usual relation (c - wave velocity):

$$\delta_r \approx \frac{c}{2\,B} \approx \begin{cases} \dfrac{1}{2}\tau_{coh}\, c & \text{for time stretched signal} \\ \dfrac{1}{2}t_w\, c & \text{for pulse shaped signal} \end{cases} \qquad (3)$$

even if it should be considered with care. The relation originates from narrowband radar whose sounding signal suffers not from signal deformation neither by reflection at small bodies nor by antenna transmission. In contrast to that, a UWB signal bouncing a point scatterer will sustain a twofold differentiation and further deformations due to the antennas. The unambiguous rage r_{ua} of the UWB radar relates to the signal repetition by:

$$r_{ua} = \frac{1}{2}t_P\, c \qquad (4)$$

Recording time: UWB sensors provide, depending on their principle of work, either the impulse response function (IRF) or the frequency response function (FRF) of the scenario under test. The time needed to collect all data for one IRF or FRF (including synchronous averaging of repetitive measurements) we call recording time T_R. Non-stationary test scenarios limit the recording time either to

$$T_R\, B_{SC} \leq \frac{1}{2} \qquad (5)$$

B_{SC} - physical (single-sided) bandwidth of the scenario variation or to

$$T_R B \leq \frac{c}{2|v|} \qquad (6)$$

$|v|$ - radial speed of a target

Equation (5) simply indicates the Nyquist theorem telling us that the refresh rate of the measurement $R = T_R^{-1}$ must be twice the bandwidth of the process to be observed. Relation (6) refers to the Doppler-effect. It is evoked from moving targets causing an expansion or compression of the scattered signal. If such signals are accumulated (by correlation or/and synchronous averaging) over a too long duration, they de-correlate resulting in an amplitude degradation of the receiving signal and finally in the loss of the target. Equation (6) should not be confused with Doppler ambiguity which is not relevant for UWB sensing.

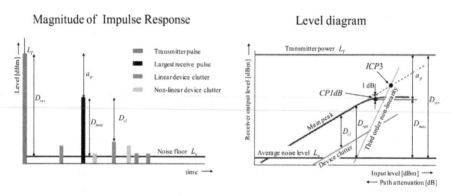

Figure 1. Stylized impulse response (left) and level diagram (right) of an UWB sensor

Dynamic range: Another group of important features relates to the sensitivity of weak signal detection. For illustration, we consider Fig. 1. The illustration on the left-hand side symbolizes the response of a single target. The red line represents the transmitter pulse (or also the auto-correlation function of a wideband CW-signal), and the black line is the target return which should be the only signal visible on the receiver screen. Obviously, we may detect many signal components hampering the detection of weak targets if there are some. These perturbing signals are random noise (electronic and quantization noise) and device internal clutter. It depends on the receiving signal and may be caused by the linear (internal mismatch, cross-coupling, frequency-dependent transmission behavior of electronic compounds) and non-linear effects (e.g. device saturation). Fig. 1 (right) depicts typical dependences of the perturbations from the level of the receiving signal. Based on this, we can derive various dynamic ranges:

- *Clutter-free dynamic range* D_{cl}: It refers to the level difference between receiving signal and the strongest internal clutter peak. D_{cl} determines the sensitivity to detect weak targets in the presence of a strong one, and also the strength of artefacts in radar images.
- *Optimum dynamic range* D_{opt} : Internal clutter caused by linear effects can be removed by sensor calibration as used in network analyzer measurements. A perfect calibration supposed, the erroneous signals are curtailed now by the noise floor and non-linear distortions. Hence, we get optimum conditions for a large dynamic range at the interception of noise and third-order distortion lines.
- *Maximum dynamic range* D_{max} : It is defined by the difference between 1 dB compression point and noise level. Its value gives a hint on the sensitivity to detect moving targets of weak reflectivity. In many cases, the strongest backscatter signals are caused by static objects. As long as the UWB sensor is not moved, these signals and their clutter contributions are stationary so that they may be simply removed by high-pass filtering in observation time. Hence, the detectability of moving targets is only limited by the noise level. The maximum dynamic range can be roughly estimated using the following relation [2]:

$$D_{max} \approx \frac{2\,\eta_r\,T_R\,V_{1dB}^2}{k\,T_0\,CF^2\,F\,R_0} = \frac{3\,\eta_r\,T_R\,B\,2^{2\,ENOB}}{CF^2} \qquad (7)$$

η_r - receiver efficiency; T_R - recording time; V_{1dB} - input voltage at 1 dB compression point (before correlation); k - Boltzmann constant; T_0 - temperature; CF - crest factor; F - noise factor; R_0 - receiver input impedance; B - receiver bandwidth; $ENOB$ - effective number of bits (before correlation).

The left part of eq. (7) applies performance parameters of analog receivers while the right part deals with the global effective number of bits merging the performance of analog and digital receiver components.

- *System performance* D_{sys} : It relates the transmitter level to the noise level. Hence, it is given by the maximum dynamic range and the attenuation of the strongest transmission path.

Time and frequency errors: While above mentioned device characteristics refer to ordinate quantities of a signal representation, the following features quantify the quality of the abscissa representation, i.e. the time or frequency axis. Related to this, we can observe systematic deviations like non-linear frequency or time axis representations resulting in non-equidistant sampling and distortions of frequency-time conversions. Random errors of the time or frequency axis representation, we call jitter or phase noise in the case of short time variation and drift for long term variations. Jitter (respectively phase noise) causes signal-dependent noise which is elevated at signal edges and disappears at flat signal parts.

Jitter limits the performance of super resolution techniques and reduces the sensor sensitivity to detect weak scattering targets in the vicinity of strong reflectors.

Efficiency: The term efficiency can be seen under different aspects. We will consider three of them here.

Receiver efficiency η_r (see also (7)): The receiver efficiency describes the capability of the receiver to exploit the incident signal energy. As the receiving signals are usually quite weak due to the restrictions of transmission power, one has to attach great importance to the receiver efficiency. It is determined by losses in the receiver front end, e.g. the insertion loss of filters or conversion loss of mixers or sampling gates. However, dead times for energy accumulation due to filter settling, incomplete data capture by reason of sub-sampling or incomplete exploitation of captured date due to serial instead of parallel data processing are much more important. Thus, the efficiency of recent UWB receivers is often reduced to values below 1 ‰ or even less which provides some potential for further improvements.

Figure of Merit FoM : In general terms, the Figure of Merit expresses the expense of energy which is required to achieve a certain effect. Two examples shall illustrate the approach. The first one deals with a Nyquist analog-to-digital converter which is aimed to digitize data with a certain rate f_s. An obvious definition of the Figure of Merit can be:

$$FoM_{flashADC} = \frac{P}{2^{ENOB} f_s} \left[\text{J/conversion} \right] \tag{8}$$

P - power dissipation of the ADC; $ENOB$ - effective number of bits of the ADC; f_s - sampling rate

Typical FoM-values for high speed ADCs are to be found at about 10 pJ/conversion. Hence, the power requirement of a 6 bit ADC @ 10 GHz is in the order of 6 W.

The second example relates to an amplifier whose FoM-value is expressed by:

$$FoM_{ampl} = \frac{P}{g \, B \, CP_{1dB}} \left[\text{Hz}^{-1} \right] \tag{9}$$

P - power dissipation of the amplifier; g - power gain in linear units; B - bandwidth; CP_{1dB} - 1 dB compression point in linear units.

The FoM-approach can be extended to further electronic components and numerical algorithms as well. We can conclude two things from FoM-philosophy. Firstly, the designer of an electronic sub-system or algorithm has to achieve a reasonable small FoM-value with his design. Secondly, the designer of the whole system gets some hints on the feasibility of his system conception and the scope of its features if the corresponding FoM-values are known.

Data throughput: UWB sensors provide lots of data particularly if they are assigned for MIMO-systems and high measurement rate. In order to conserve energy, memory space and data transmission capacity, the sensors should not provide unnecessary data. We have six basic options to reduce the data throughput:

- The data should be captured close to the Nyquist rate.
- The length of the measured impulse response should not be much longer than the settling time of the scenario under test.
- Synchronous averaging (if appropriate) should be performed immediately after data capture.
- A short word length of digitized data should be kept by avoiding high crest factor signals.
- Stationary data should be removed by feedback sampling or digital filtering immediately after data capture (see also chapter 3.3.4 in [2]), and
- Sparse or compressive sampling [3] should be performed. However, this point will not be considered here as it would go beyond the scope of this chapter.

Without going into detail, we would like to mention at least some further aspects that influence the performance of sensor operation, too. They concern interference issues like robustness against jamming and low probability of intercept (LPIR- low probability of intercept radar).

The performance figures summarized above are the basis for deciding on a certain sensor configuration for a specific application. In what follows, the most popular UWB sensor principles will be tabulated and assessed with respect to the introduced performance figures.

2.2. Principles of UWB-sensors

We divide the UWB sensor principles into two groups. While the sensors of the first group generate sounding signals of large instantaneous bandwidth, the devices belonging to the second group deal with narrowband signals swept over a large bandwidth. A thorough analysis of the different sensor concepts of both groups including a reference list can be found in [2]. Here, we will only give a short summary to get an impression of the most common sub-components of UWB sensors and to understand the advantages and disadvantages of the various principles.

2.2.1. Sensors of large instantaneous bandwidth

There are several UWB approaches known exploiting signals of large instantaneous bandwidth. Usually, they are denoted according the sounding signal applied by the sensor. Typical representatives of this signal class are:

- sub-nanosecond pulses
- very wideband pseudo-noise codes

- multi-carrier signals (also assigned as multi-sine), and
- white random noise.

By assumption, these signals have a bandwidth in the GHz range requiring often Nyquist rates of the measurement receivers above 10 GHz. Disregarding the device costs, this is hardly to achieve with the limited power budget and the restricted means of data handling (see section 2.1 – *Figure of Merit* and *Data throughput*) which a sensor usually has at its disposal. Hence, all these devices must reduce their data rates at the expense of receiver efficiency, which is reflected by a reduced dynamic range D_{max} (see(7)). The data rate reduction is either achieved by sub-sampling or by serializing the data recording.

Fig. 2 refers to three possible device conceptions for illustration. The two upper approaches require periodic sounding signals. Here, the signal shaper may be a pulse generator, a binary PN-generator or an arbitrary waveform generator. The most often found device implementations apply sub-nanosecond pulse generators. Indeed, the concept allows the implementation of very cost-effective and power saving sensors. However, their system performance often suffers from reduced dynamic range due to the large crest factor of the sounding signal (compare (7)); they do not provide jitter suppression (see also sub-chapter 2.3) and they are not robust against jamming. Wideband PN-generators are an interesting alternative to pulse generators since they provide powerful signals of low magnitude (i.e. of low crest factor). Arbitrary waveform generators are able to provide signals which can flexibly be adapted to the measurement problem. However, they are quite expensive, power hungry and limited with respect to the bandwidth. Hence, they have not been found in practically applicable sensor concepts recently.

Sub-sampling receiver: It is the most often applied UWB concept. It supposes periodic sounding signals (t_P - signal period). Typically, the measurement signals are captured by sequential sampling, providing one data sample per period whose time position is stepwise shifted over the whole signal. The actual sampling interval is $t_P + \Delta t$, while the equivalent sampling interval which has to meet the Nyquist criteria is Δt. Newer concepts apply interleaved sampling permitting higher sampling rates since more than one point per period is taken. The classical concept of time shift control uses the fast ramp-slow ramp approach which, however, tends to non-linear time axis representation, sampling jitter and time drift. A second method deals with two stable sine wave generators (e.g. Direct Digital Synthesizers of slightly different frequency ($f_1 = t_P^{-1}$; $f_2 = (t_P + \Delta t)^{-1}$). This reduces time drift and avoids time axis non-linearity. However, it still keeps the sampling jitter quite high since the trigger events launching the sounding pulse and activating the sampling gates are based on relative flat edges of the two sine waves of (comparatively low) frequency f_1 and f_2. Timing control based on digital counters for coarse timing exploits steep trigger edges improving the jitter performance. Then, the fine tuning is typically done by programmable delay chips which consist of hundreds of delay gates. As these gates are not absolutely identical, the delay line cannot ensure equidistant sampling.

Furthermore, the delay time depends on temperature, and the huge number of gates consumes plenty of energy.

Sub-sampling receiver

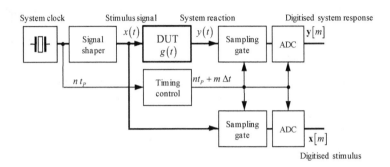

Analog Correlator

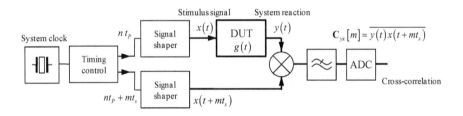

Sub-sampling correlator

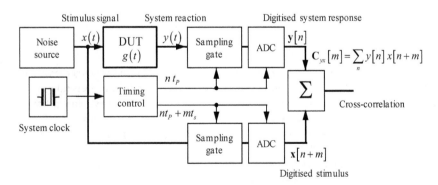

Figure 2. Three possible sensor structures exploiting signals of large instantaneous bandwidth.

Analog correlator: Due to the lag of programmable analog wideband delay lines, one applies two wideband sources (pulse or PN-sequence) providing two identical signals which are shifted in time. The time shift may be controlled by the same approaches as mentioned above. One of these signals stimulates the DUT, and the other one acts as reference in a correlator. Even if the mixer and the integrator do not waste signal energy, the correlator has about the same efficiency as a sequential sampling receiver as long as one does not deal with parallel correlation stages. We can find from eq. (7) that the correlation principle will provide the best dynamic range due to the large time-bandwidth product. But this benefit will be gambled away if sounding signals of large crest factors are applied.

Sub-sampling correlator: Here, we can use also random noise as stimulus. The time lag between measurement and reference signal is performed by shifting the sampling time as explained before. The correlation is done in the numerical domain. The approach is quite time consuming since the averaging time must be high in order to achieve a stable estimation.

2.2.2. Sensors of narrow instantaneous bandwidth

Strictly spoken, such sensors do not belong to UWB systems but they are doing the same job as real UWB devices if they are applied for sensing. Hence, they are worth being considered.

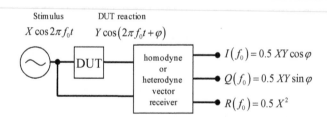

Figure 3. Principle of sine wave measurement.

Fig. 3 depicts the very basic principle. The signal source is a sine wave generator which steps or sweeps the signal frequency over the wanted bandwidth. Depending on the requirements (signal purity, frequency stability, frequency axis linearity, settling time etc.), free-running VCO's, synthesizers or DDS-circuits are in use. The receivers are based on homodyne or heterodyne down-conversion providing the complex valued frequency response function of the DUT:

$$\underline{G}(f) = \frac{I(f) + jQ(f)}{R(f)} \xrightarrow{\;IFFT\;} g(t) \tag{10}$$

which can be transformed via IFFT into the impulse response function. Simple implementations (e.g. many FMCW-radars) abstain from vector receivers. They only deal with the in-phase component.

Measurement principles applying sine waves provide the best suppression of noise and harmonic distortions due to narrowband filtering before signal capture. Their receiver efficiency tends to one as long as the settlement of resolution filters and signal source are negligible against the recording time. Hence, such devices often suffer from long measurement duration which leads to a strong range-Doppler coupling. The recording time can be reduced either by simultaneous measurements at different frequencies [7] (requiring complex parallel receiver and synthesizer) or by renouncing the narrowband filters (giving up the sensitivity benefits compared to the wideband approaches).

2.3. UWB pseudo-Noise Concept

Under the assumption of Pseudo-Noise (PN)-codes for sounding, Nyquist sampling for data capture and embedded pre-processing for data reduction, the principle depicted on the top of Fig. 2 seems to be the most promising if one trades the pros and cons of the various UWB principles with respect to monolithic integration, system performance, MIMO-capability and power consumption. Fig. 4 represents the modified structure adapted to the conditions mentioned above. The use of two receiver channels yields the best performance with respect to different application aspects like synchronous measurement of stimulus and reaction signal, opportunity of device calibration, difference or interferometric measurements as well as long term sensor stability.

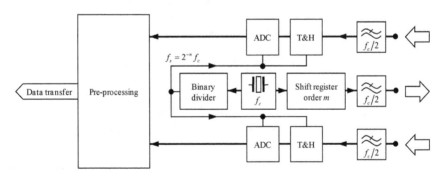

Figure 4. Basic structure of UWB PN-device.

A stable microwave oscillator controls the whole system. It has to provide only a single frequency f_c which allows the use of simple and stable generator concepts. The oscillator pushes a high-speed shift register. Depending on its feedback structure, it provides any binary sequence. Preferentially, M-sequences are used due to their favorable auto-correlation function. Other options could be Golay-codes [8] or Gold-codes if cross-correlation properties are in the foreground of interest.

For the sake of brevity, we restrict ourselves here solely to M-sequences. These codes have a sinc²-spectrum with the first zero located at f_c and they have concentrated about 80% of their energy in the spectral band from DC to $f_c/2$. Hence, we will not provoke a dramatic performance loss if we limit the upper frequency to $f_c/2$. On the contrary, the band limitation avoids disproportionate growth of noise against ever decreasing signal power.

The band limitation to half the clock rate touches several performance-relevant issues:

1. We need a sampling rate of at least f_c in order to meet the Nyquist theorem. In other words, it is sufficient to capture one sample per chip of the M-sequence. As the M-sequence is periodic, we can do this by sub-sampling. It is easy to show that a binary divider is sufficient for timing control since the number N of chips in the sequence is always one less than a power of two ($N = 2^m - 1$). If the order m of the shift register and the order n of the binary divider are identical, we have sequential sampling. For $n < m$, one speaks about interleaved sampling which takes more than one sample within one period.

2. Both, the signal edges of the microwave clock f_c as well as of the sampling clock f_s are quite steep. Hence, the trigger events activated by them are robust against jitter and drift.

3. The time axis of the receiver is defined by the sampling clock f_s. This clock originates from a stable RF-generator and a digital frequency divider which has to run trough all its states before it can launch a new impulse. Hence, any internal deviations between the involved flip-flops have no effect on the divided signal. Therefore, apart from the remaining jitter, we can expect exact equidistant sampling i.e. an absolutely linear time axis representation.

4. The principle of interleaved sampling allows the sampling rate to be varied by keeping the sensor concept. Thus, one can reduce the sampling rate in favor of reduced power consumption and device costs or it can also be increased to improve the receiver efficiency η_r depending on the development state of high-speed electronics.

5. Nyquist sampling provides the lowest possible data throughput[1] without violation of sampling theorem.

The embedded pre-processing is mainly aimed at data reduction by synchronous averaging (often the measurement rate is much higher than required by the time variance of the test scenarios), static background removal or signal transformations. It should, however, be respected that impulse compression (in order to get the impulse response) performed at this point will increase the data throughput toward the main processor since the word length of the data samples increases.

---

[1] Here, we refer to general measurement conditions. We disregard sparse sampling which is largely dependent on the measurement objects.

The sensor principle depicted in Fig. 4 is basically also able to deal with short sub-nanosecond pulses. However, this would greatly degrade the performance of the system which is largely determined by the amount of signal energy accumulated in the receiver. In the case of pulse signals, this requires amplifiers of high compression points and high resolution ADCs since the whole signal energy is concentrated in a short moment. Furthermore, the measurement object may be exposed to strong fields in the case of near-field measurements. The application of PN-codes avoids all these flaws since it carries enough energy even with small signal magnitudes. As the impulse compression (leading to high crest factor signals) is performed in the digital domain, the analog sensor components and test objects are spared from high voltage peaks.

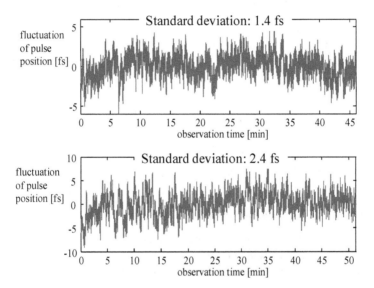

Figure 5. Stability of time position measurement. The delay time of a 50 cm long RF cable was measured at constant temperature. The RF clock was provided either by a sophisticated sine wave generator (SMP04 from Rhode & Schwarz) (top) or by a free running DRO (bottom).

It is well known that the impulse compression of time-extended wideband signals largely improves the dynamic range. As shown in [2] (chapter 4.7.3), it also reduces the jitter susceptibility. The impulse compression distributes the jitter power evenly over the whole signal like additive noise. However, a noise increase above the "natural" level of electronic and quantization noise cannot be observed since the jitter-induced perturbations remain quite low due to the measures described above. Hence, the edges of the impulse response of a DUT measured by the PN-principle are not affected by jitter as usually in pulse measurements. This favors the PN-sensor concept for applications dealing with super resolution techniques or micro-Doppler problems, particularly if weak scattering targets are overwhelmed by strong ones. Chapter11 *ultraMedis* gives some examples of related problems, and Fig. 5 illustrates the achieved short-time stability of an M-sequence sensor having a bandwidth of about 8 GHz. The short-time variance of the pulse position measurement was in the lower fs range corresponding to a distance variation below 1 μm.

The simple timing concept of the PN-sensors enables the implementation of large MIMO-arrays at which the number of cascaded measurement units is basically not limited. The principle is shown in Fig. 6. However, the data handling will be increasingly demanding with a rising number of channels. In a typical operation mode, the transmitters are sequentially activated while the receivers of all channels work in parallel. Some details of implemented MIMO-systems can be found in chapter 11 *ultraMedis*.

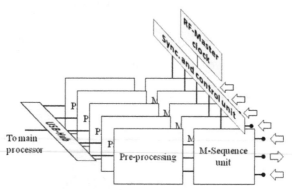

Figure 6. Creation of a MIMO-system by cascading M-sequence sensors.

The receiver of the UWB PN-sensor applies sub-sampling for data capture. Hence, its efficiency gives some potential of further improvements. This would, however, be connected with a considerable increase of the sampling rate f_s. As we can see from (8), the elevation of the sampling rate has to be done at the expense of the ADC resolution since the FoM-value is primarily fixed by the semi-conductor technology, while the maximum power is limited by the achievable heat transport. However, simply increasing the sampling rate based on low bit ADCs will not bring any profit with respect to the sensor performance, i.e. the opposite will happen.

As, however, the update rate of UWB PN-sensors is much higher than required by the time variance of the test object, the difference between two consecutive measurements is very low so that low resolution ADCs are sufficient for capturing these deviations. Anyway, this supposes a fast control loop and a (less power hungry) DAC of sufficient resolution which provides the captured signals from previous measurements for reference. Some basic considerations related to this type of feedback sampling can be found in [2]. Details of the layout and implementation of related sub-components are discussed in sections 3.4, 5 and 6.2.

3. Analog wideband receiver circuits

3.1. Introduction

In order to support system design in its individual stages, different amplifier versions have been developed. Classical ultra-wideband low-noise amplifiers (UWB-LNAs) have been implemented first to ensure early availability and to assess the SiGe BiCMOS technology applied. Then, new receiving components have been considered to address the requirements discovered in system design. This way, a new subtraction amplifier has been made available which allows for practical evaluation of the feedback sampling approach. **Fig. 7** illustrates the way in which LNAs and subtraction amplifiers are used as part of the system.

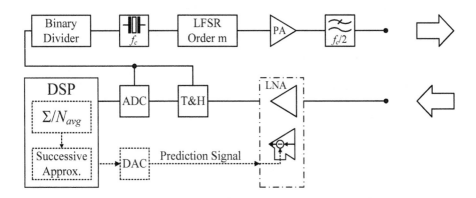

Figure 7. Transmit and receive path of the basic UWB PN-device from Fig. 4 (solid lines) and proposed extension to allow for feedback sampling (dashed lines – see subsection 6.2.3). Either the LNA component or the amplifier with integrated signal subtraction capability are present (dash-dotted box) according to the system state at which it is focused.

In this figure, the basic M-Sequence system is indicated by solid lines. Within this system, classical UWB-LNAs (the LNA component in the dash-dotted box) are used to support the way of operation presented in subsection 2.3. Possible implementations are covered in subsections 3.2.1 to 3.2.3. In addition, a system extension is indicated by the dashed lines, which tries to overcome the limitations imposed by the analog-to-digital converter. For this reason, a digitally calculated prediction signal is provided by a digital-to-analog converter and subtracted from the receive signal to generate a difference signal of highly reduced signal swing close to sampling time instances (see sections 5.1 and 6.2.3). Subtraction is performed by an amplifier with integrated signal subtraction capability, practical realizations of which are presented in sections 3.4.1 and 3.4.2.

The latest subtractor also accounts for low feeding point impedance, which is imperative for the conceptual design of dielectrically scaled antennas. The use of such a device is intended by the collaborative project *ultraMedis* (see chapter 11). For establishing common interface definitions, the performance of individual components has to be characterized by appropriate metrics. While those are well established for single-ended arrangements, this is not the case with the noise characterization of multiport or differential structures. Hence, a new de-embedding scheme for the noise figure of a differential device has been developed and will be presented.

3.2. LNAs for the basic M-Sequence system

Within the basic M-sequence system, low noise amplifiers (LNAs) perform the classical task of adapting the input signal swing to the dynamic range of the analog-to-digital converter (ADC) while adding only a minimum amount of excess noise and providing reasonable power-match conditions. If a high-gain LNA is used, the system also is less sensitive to noise added by succeeding components. Gain, in turn, is limited by the required linearity, and an appropriate compromise with respect to all counteracting requirements has to be found. While trading one parameter against the other, the conditions set by the technology have to be considered for the individual LNA. Resonant tuning and resistive feedback topologies are predominantly used in literature for mapping specifications to circuit designs. Though a resonant solution is favored in [11], the authors do admit that the parasitic base resistance of bipolar transistors causes a large contribution to output noise. Thus, the advantage of extraordinary low noise figures enabled by narrow-band resonant designs as compared to designs matched by resistive feedback is relativized. High magnetic field gradients potentially encountered in some of the applications, and the limited ability to use shielding as identified in [12] also make the use of inductors questionable. Therefore, resistive feedback solutions have been preferred as their use is additionally accompanied by notable die size advantages. While the design of individual amplifiers will be covered in the following subsections, general guidelines can be taken from standard textbooks. In [13], for example, the impact of feedback on noise and impedance match is analyzed in detail. For the design, too, a simplified version of the bipolar transistor small signal equivalent circuit model with additional noise sources as presented in [13] has been used.

3.2.1. Multiple resistive feedback LNA

One of the implemented amplifier versions which have been inspired by classical UWB-LNAs is depicted in Fig. 8. According to [10], this is a popular wideband amplifier topology often referred to as *Kukielka* amplifier. Due to numerous results reported in literature for this kind of amplifier, the impact of technology on circuit performance can be assessed. For comparison, especially SiGe implementations as presented in [9] are valuable. The main characteristics of this amplifier are set by the core circuit which comprises transistors Q_1 to Q_3. For analysis, the Darlington pair $Q_2 - Q_3$, which is used for gain-bandwidth extension of the second stage, is treated as single compound-transistor Q_{23}. Within the simplified circuit thus obtained, four feedback loops can be identified. For proper biasing, series-series feedback is applied to Q_1 and Q_{23}. By this measure, the bandwidth of both stages is improved. In turn, input and output impedances of the amplifier are increased rather than decreased as required for input and output power matching.

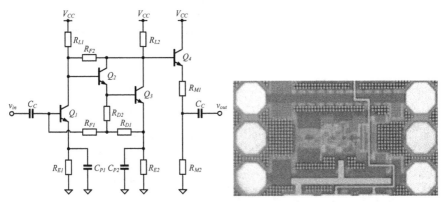

Figure 8. Schematic diagram (left) and chip photo (right) of the multiple resistive feedback LNA - the die area is only 530 μm × 280 μm.

Hence, local shunt-shunt feedback is applied to Q_{23} in order to reduce the output impedance. Finally, to enable input power matching, global shunt-series feedback from the emitter node of Q_{23} to the base of Q_1 is applied. According to [10], this configuration tends to present an overdamped response. For this reason, peaking capacitors C_{P1} and C_{P2} are inserted to improve the frequency behavior of the amplifier. The addition of peaking capacitors might, however, impair stability which has to be diligently observed during design for this reason as stated in [13]. In the same publication, an approximate calculation of the noise figure (NF) for this topology reveals that the latter is dominated by the noise properties of Q_1 as long as $R_{F1} \gg R_S$ and $R_{F1} \gg R_{eq}$. R_{eq} is related to the equivalent input noise voltage source v_{Q1} of transistor Q_1 by $R_{eq} = v_{Q1}^2 / (4kT)$. For this reason, the selection of transistor Q_1's bias current to yield optimal current density with respect to its noise properties should be a first step in design. After this initial step, one of the directed design procedures given in [10] or [9], respectively, can be used for further development. Those are derived from input and output power match conditions as well as from pole positions. In order to account for the characteristics of the Darlington pair transistors Q_2 and Q_3, substitutions $g_{m,23} \to g_{m,3}$ and $\omega_{T,23} \to 2 \cdot \omega_{T,3}$ can be applied according to [10]. In Fig. 8 (left), an emitter follower has been attached for further improving the output power matching in the technology used. Fig. 8 (right) shows the chip photograph of the implemented LNA without pads used for biasing. The dimensions of the displayed die area are $530\,\mu m \times 280\,\mu m$ only, which confirms the advantage resistive feedback amplifiers provide in view of die area as compared to resonant solutions. For the accurate characterization of the fabricated amplifier, on-waver measurements have been performed using a PM 8 probe station of Süss MicroTec (now acquired by Cascade Microtech). Due to the measurement arrangement, losses preceding and succeeding the device under test (DUT), i.e. the amplifier, cannot be avoided. However, their impact on the scattering parameters of the DUT can be eliminated by proper calibration of the network analyzer applied. Also, the spectrum analyzer with noise figure measurement personality in use allows for the specification of losses preceding and succeeding the DUT which are compensated for during measurement in this case[2]. Measurement results obtained in this way are shown in Fig. 9 together with results from post-layout simulation. Initially, they have been presented in [14].

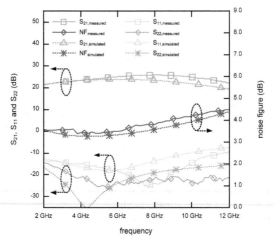

Figure 9. Measurement results of the multiple resistive feedback LNA. (The arrows point to the related axis scaling.)

From Fig. 9, peaking in the gain curve progression of the measurement data can be observed as compared to the results from post layout simulation. The maximum difference appears at about 8 GHz, which is the frequency at which a notch in measured S_{11} values also appears. In [14] it is thus suspected that this deviation arises due to the interaction of the test set-up with the DUT. In short summary, the results presented in Fig. 9 for the low-cost technology applied map pretty well the state-of-the-art performances reported in literature at that time. For a more detailed analysis, the reader may consult [14].

3.2.2. Active Feedback LNA

This amplifier has been inspired by the work presented in [15]. Due to the characteristics of the applied technology, certain adaptations have been required, though. Fig. 10 (left) shows the schematic diagram of the final design.

---

[2] Rhode & Schwarz ZVA-24 and FSQ-40 have been used for the measurements.

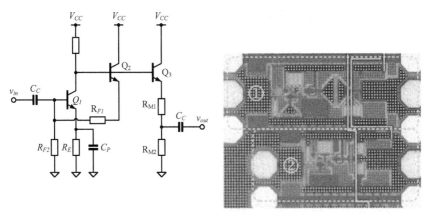

Figure 10. Schematic diagram (left) and chip photo of both active feedback LNA versions (right). The die areas are (1) $630\,\mu m \times 280\,\mu m$ and (2) $530\,\mu m \times 280\,\mu m$, respectively.

It is a one-stage amplifier with resistive emitter degeneration to ensure a stable DC operating point and to improve bandwidth. Due to the presence of the peaking capacitor C_P, degeneration is continuously shifted to higher frequencies. As in the case of the multiple resistive feedback amplifier, this technique has to be used with care to ensure that this measure does not impair amplifier stability. Input matching of Q_1 is achieved by feedback via transistor Q_2 as well as resistors R_{F1} and R_{F2}. The advantage of Q_2 is twofold: It improves the isolation between the input node and the output node in forward direction and, according to [15], it helps to enlarge the collector-emitter voltage of Q_1. Thus, the maximum oscillation frequency f_{max} is expected to be increased, and the large-signal behavior is said to be improved. The amplifier according to Fig. 10 was presented in [14] for the first time. Compared to the amplifier in [15], the inductor used for improving the frequency behavior has been abscised from the design while the peaking capacitor has been added. Also, for better output matching, emitter follower Q_3 has been attached. To avoid a lengthy discussion of circuit characteristics, Fig. 10 (left) uses an alternative way to depict the circuit as compared to [14] or [15]. This representation points out the large similarity of feedback paths in both amplifiers shown in Fig. 8 and Fig. 10. Though the actual implementations of the passive feedback networks differ, shunt-series feedback is applied for input matching in both cases, and many design steps can be executed by analogy. Fig. 10 (right) shows the chip photograph of two variants, which have been implemented to assess the impact of layout on circuit performance. In the first version, 90° lead corners are avoided and consecutive 45° lead corners are used instead. By this measure, the average lead length is increased. By contrast, a compact design has been targeted in the second layout version. As discussed in [14], results do not differ significantly as long as the length of the lead connecting the RF input with the first amplifying transistor is kept comparably long. Thus, only results for the second layout version are shown in Fig. 11.

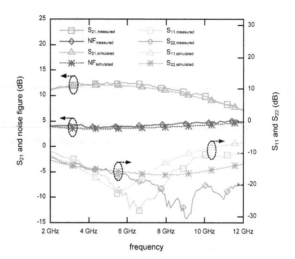

Figure 11. Measurement results of the active feedback LNA. (The arrows point to the related axis scaling.)

The die area occupied by the first layout version is $630\mu m \times 280\mu m$ while the second layout version only occupies $530\mu m \times 280\mu m$ (both excluding DC-pads). Measurements have been performed the same way as explained for the multiple resistive feedback LNA. Compared to Fig. 9, gain is much lower, which is expected due to the single stage nature of the active feedback amplifier. At the same time, the input referred $1dB$ compression point is improved notably. A more complete discussion of amplifier characteristics is presented in [14].

3.2.3. Pseudo-differential LNA

Core of the half-circuit shown in Fig. 12 (left) is the cascode amplifier with reactive shunt feedback on the left-hand side. As for the single-ended amplifiers, bias current of this arrangement should be selected due to noise and linearity considerations. While a more detailed analysis of this topology, as presented for the inductively degenerated cascade amplifier with capacitive shunt feedback in [16], might be desirable at this point, limited space for this section does not permit a lengthy discussion. Instead, we allude to the amplifier of [17] from which the topology of Fig. 12 (left) has been derived.

A pseudo-differential amplifier has been implemented to support the development of the basic M-sequence system by adding the capability to use differential circuitry, especially differential antennas. Many of its characteristics are inherited from the topology of [17]. However, emphasis with respect to certain design parameters has been shifted. The most peculiar aspect is the fact that the input matching network used in [17] could be abscised from the design. This modification was enabled by improved input-output isolation due to altered feedback tapping points. Some results of the manufactured chip, a photo of which is contained in Fig. 12 (right), are summarized in Fig. 13. Together with additional topological aspects, they are discussed in [18] in more detail. Measurements to gather those results have been performed on-waver at the PM 8 probe station using ground-signal-signal-ground (GSSG) probes. Similar to the single-ended case, scattering parameters could be determined directly by a calibrated network analyzer. By contrast, only single-ended equipment has been available for noise figure measurement, and it is left to the next subsection to discuss a method applicable to (pseudo-)differential amplifiers.

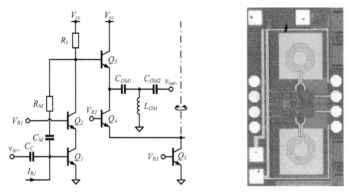

Figure 12. Half-circuit schematic diagram (left) and chip photo (right) of the pseudo-differential LNA.

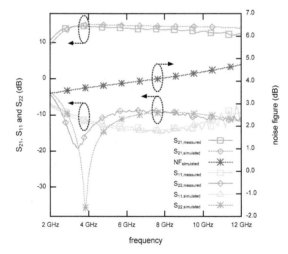

Figure 13. Measurement results of the pseudo-differential LNA. (The arrows point to the related axis scaling.)

3.3. Differential noise figure measurement

Due to space limitations, only a short introduction to this topic will be presented here. While there are alternative methods as presented in [23] and [24], for example, focus will be on a new powerful method proposed in [20]. A convenient definition for the noise figure of a differential (or multiport) device with respect to one of its ports has been given by Randa [22] and is reprinted in (11) with slight modifications:

$$F_i\left(\mathbf{\Gamma}_k, \mathbf{C}_k\right) = \frac{\left\{\left[\mathbf{I} - \mathbf{S}_k\mathbf{\Gamma}_k\right]^{-1}\left(\mathbf{S}_k\mathbf{C}_k\mathbf{S}_k^\dagger + \mathbf{C}_{S,a}\right)\left[\left[\mathbf{I} - \mathbf{S}_k\mathbf{\Gamma}_k\right]^{-1}\right]^\dagger\right\}_{ii}}{\left\{\left[\mathbf{I} - \mathbf{S}_k\mathbf{\Gamma}_k\right]^{-1}\mathbf{S}_k\mathbf{C}_k\mathbf{S}_k^\dagger\left[\left[\mathbf{I} - \mathbf{S}_k\mathbf{\Gamma}_k\right]^{-1}\right]^\dagger\right\}_{ii}} \tag{11}$$

The noise figure in (11) is parameterized by the matrix of reflection coefficients seen by the DUT into the ports of connected components $\mathbf{\Gamma}_k$ and the noise correlation matrix of incident noise waves injected by an external source $\mathbf{C}_k$. $\mathbf{I}$ is the identity matrix in (11), the dagger indicates the Hermitian conjugate, and $\mathbf{S}_k$ as well as $\mathbf{C}_{S,a}$ are the scattering matrix and the noise correlation matrix of emergent waves contributed by the DUT, respectively. To apply this definition, $\mathbf{C}_{S,a}$ has to be determined first. Therefore, the differential device has to be embedded into a network of passive components which provide the differential excitation as only single-ended measurement equipment is currently available. This is demonstrated in Fig. 14. The noise correlation matrix of the DUT then has to be de-embedded from the results measured for the component chain.

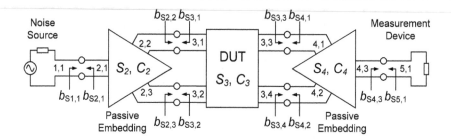

Figure 14. DUT embedded into passive components. $\mathbf{b}_{Sx,y}$ are equivalent generator noise waves caused by component x and ejected from the respective port y.

For this purpose, the noise distribution matrix defined in [21] is a convenient starting point. Multiplied by Boltzmann's constant k and the physical temperature T, it is the correlation matrix of emergent noise waves caused by a passive component which account for all the noise generated within the device. It can be related to the scattering matrix as shown by (12). A short and intuitive proof of this relation is contained in [25]:

$$\hat{N} = \left(\overline{\mathbf{b}_s \mathbf{b}_S^\dagger} \right) / (kT) = \mathbf{I} - \mathbf{SS}^\dagger \tag{12}$$

Making use of single-ended noise measurement equipment, the characteristic noise *equation reference goes here*temperature $\hat{T}_{chain}$, which characterizes noise from all the elements of the whole component chain, can also be determined. Thus, we are only left with the problem to describe signal transfer via the component chain to accomplish the goal of de-embedding $\mathbf{C}_{S,a}$. In [20], an approach based on the connection scattering matrix $\mathbf{W}$ has been presented for this purpose, which surpassed the method of [19], because it is applicable without simplifying assumptions. The connection scattering matrix was introduced long ago, and its use for computer-aided circuit analysis has been discussed in [26], for example. For all ports in a component network, it relates the incident to the impressed waves. To enable such matrix representation, incident, reflected, and impressed waves have to be composed to wave vectors $\mathbf{a}$, $\mathbf{b}$, and $\mathbf{b}_S$, respectively, which should be sorted in a component-wise way. For convenience, a component index is assigned to the vector entries, and it is assumed that elements corresponding to the DUT (index k) are placed at the bottom of each wave vector. An element is thus identified by two indices i, j representing the component and the respective port number. With this convention, all complex wave amplitudes can be related by the set of linear equations

$$\mathbf{b} = \mathbf{S} \cdot \mathbf{a} + \mathbf{b}_S \tag{13}$$

where $\mathbf{S}$ is a block-diagonal matrix assembled from the individual component S-parameter matrices. The connections between the single components impose additional constraints on the wave amplitudes. To account for them, the connection matrix $\mathbf{\Gamma}$ is used as in (14):

$$\mathbf{b} = \mathbf{\Gamma} \cdot \mathbf{a} \tag{14}$$

Most often, a common real reference impedance is applied for all components. Then, all entries of $\mathbf{\Gamma}$ are zero except those which refer to connected ports and, thus, are one. In this case, $\mathbf{\Gamma}$ is a permutation matrix with $\mathbf{\Gamma}^{-1} = \mathbf{\Gamma}^T$. From (13) and (14), the incident wave vector $\mathbf{a}$ can be eliminated to get

$$\mathbf{W} \mathbf{\Gamma}^{-1} \mathbf{b} = \mathbf{b}_S, \; with \; \mathbf{W} = \mathbf{\Gamma} - \mathbf{S} \; \mathbf{W} = \mathbf{\Gamma} - \mathbf{S} \tag{15}$$

Noise generated by the source does not contribute to the characteristic noise temperature of the component chain. The respective entries of $\mathbf{b}_S$ can thus be set to zero. Furthermore, in the case of $\mathbf{S}_1$ containing zero entries[3], the corresponding rows of $\mathbf{W} \mathbf{\Gamma}^{-1}$ should be deleted to avoid rank deficient matrix problems in some of the computations. For convenience, we will refer to the matrix obtained from this operation by $\mathbf{V}$. After additional matrix partitioning, which is required later, (15) then becomes

---

[3] $\mathbf{S}_1$ is a submatrix of $\mathbf{S}$.

$$
\mathbf{b}_{S,p} \begin{bmatrix} \mathbf{b}_{S,2} \\ \vdots \\ \mathbf{b}_{S,k-1} \\ \mathbf{b}_{S,k+1} \\ \vdots \\ \mathbf{b}_{S,n} \\ \mathbf{b}_{S,k} \end{bmatrix} = \begin{bmatrix} \mathbf{V}_{pm} & \mathbf{V}_{pp} & \mathbf{V}_{pa} \\ \mathbf{V}_{am} & \mathbf{V}_{ap} & \mathbf{V}_{aa} \end{bmatrix} \begin{bmatrix} \mathbf{b}_1 & \mathbf{b}_m \\ \mathbf{b}_2 \\ \vdots \\ \mathbf{b}_{k-1} \\ \mathbf{b}_{k+1} & \mathbf{b}_p \\ \vdots \\ \mathbf{b}_n \\ \mathbf{b}_k & \mathbf{b}_a \end{bmatrix} \tag{16}
$$

In (16), submatrix $\mathbf{V}_{am}$ will always be zero, because there is no direct connection from the input to the DUT. Hence, we are left with

$$
\mathbf{b}_{S,p} = \mathbf{V}_{pm}\mathbf{b}_m + \begin{bmatrix} \mathbf{V}_{pp} \mid \mathbf{V}_{pa} \end{bmatrix} \begin{bmatrix} \mathbf{b}_p \\ \mathbf{b}_a \end{bmatrix}, \tag{17}
$$

and

$$
\mathbf{b}_{S,a} = \begin{bmatrix} \mathbf{V}_{ap} \mid \mathbf{V}_{aa} \end{bmatrix} \begin{bmatrix} \mathbf{b}_p \\ \mathbf{b}_a \end{bmatrix}. \tag{18}
$$

Setting $\mathbf{Q} = \begin{bmatrix} \mathbf{V}_{ap} \mid \mathbf{V}_{aa} \end{bmatrix} \begin{bmatrix} \mathbf{V}_{pp} \mid \mathbf{V}_{pa} \end{bmatrix}^{-1}$ for abbreviation, (19) is obtained from (17) and (18) after some algebra[4]:

$$
\mathbf{Q}\mathbf{V}_{pm}\mathbf{b}_m = \mathbf{Q}\mathbf{b}_{S,p} - \mathbf{b}_{S,a}. \tag{19}
$$

Finally, since $\mathbf{b}_{S,p}$ and $\mathbf{b}_{S,a}$ arise from different sources and are thus uncorrelated, $\mathbf{C}_{S,a}$ can be determined from (20)

$$
\mathbf{Q}\mathbf{V}_{pm}\mathbf{C}_m\mathbf{V}_{pm}^\dagger\mathbf{Q}^\dagger = \mathbf{Q}\mathbf{C}_{S,p}\mathbf{Q}^\dagger + \mathbf{C}_{S,a} \tag{20}
$$

with $\mathbf{C}_m = \overline{\mathbf{b}_m\mathbf{b}_m^\dagger}$, $\mathbf{C}_{S,p} = \overline{\mathbf{b}_{S,p}\mathbf{b}_{S,p}^\dagger}$, and $\mathbf{C}_{S,a} = \overline{\mathbf{b}_{S,a}\mathbf{b}_{S,a}^\dagger}$.

In (20), noise correlation matrices of single passive components given by the product of kT and (12) are composed to the block diagonal matrix[5] $\mathbf{C}_{S,p}$. $\mathbf{C}_m$ accounts for noise from all

---

[4] As the right division function of a math program can be used to compute $\mathbf{Q}$, there is no need for an explicit inversion, and a minimum norm solution is obtained for a non-square system.

[5] Note that noise contributions of the output loads, according to [22], should not be considered for NF computation. With respect to the set-up of Fig. 14 this implies $\overline{b_{S,51} \, b_{S,51}^*} = 0$, which can be achieved by using the second stage correction of the NF meter.

components of the chain related back to the input. For the set-up of Fig. 14, $\mathbf{C}_m$ is a 1×1 matrix associated with the characteristic noise temperature as shown by (21):

$$\mathbf{C}_m \approx \left[k \hat{T}_{chain} \right] \qquad (21)$$

For using (11) in noise figure computations, its parameter matrices Γ_k and C_k still need to be determined. Γ_k contains reflection coefficients, which relate waves injected from the DUT $b_{S,a}$ to those reflected back from the embedding network a_a. It follows from inspection that Γ_k is a submatrix of $\mathbf{W}^{-1}$. Focusing on C_k, a simple argument leads to some confidence that (22) is a reasonable choice: Assume that the power splitter at the input of Fig. 14 only excites the differential mode. In this case, noise at the input of the DUT is completely correlated. Also, there should be a noise power of kT available at the input of the DUT in differential mode to stay comparable to the standard noise figure definition. Then, due to the properties of mixed-mode transformation, (22) is the evident solution. This is discussed in [20] in more detail.

$$\mathbf{C}_k = \begin{bmatrix} 0.5kT & -0.5kT & 0 & 0 \\ -0.5kT & 0.5kT & 0 & 0 \\ 0 & 0 & 0 & 0 \\ 0 & 0 & 0 & 0 \end{bmatrix} \qquad (22)$$

Thus, all inputs required for (11) are determined, and the noise figure with respect to a certain output port can be calculated. Instead of a physical port, also a logical port can be considered. For this purpose, matrices in the numerator and the denominator of (11) have to be transformed by an appropriate transformation. In view of Fig. 14, mixed-mode transformation provides noise power spectral densities in the differential mode at the selected output of the DUT for both cases: Noise generated by the DUT and the input sources, as well as noise generated by the input sources alone. Their ratio finally determines the differential noise figure of the device. This approach has been applied to the pseudo-differential amplifier shown in Fig. 15. The result is contained in Fig. 16 together with the noise figure measured from one signal branch, which is given for comparison. In the measurements, the losses of the probe heads have been appropriately taken into account.

3.4. Solutions for the feedback-sampling approach

The introduction of the feedback-sampling concept by system design spawned the requirement of signal subtraction at the input of the receive path. Hence, the demand for new components equipped with two inputs arose - one for the RF signal taken from the antenna, and another one for a digital prediction signal provided by the signal processing via a digital-to-analog converter (DAC). In theory, subtraction results in an output signal of highly reduced voltage swing to which the analog-to-digital converter (ADC) used for signal acquisition will be exposed. To confirm this theory in practice, two versions of an input subtractor have been implemented and will be presented in the following subsections.

3.4.1. Pseudo-differential feedback-sampling amplifier

Similar to the design of amplifiers for the basic system, the development of the feedback-sampling amplifier has been guided by the assumption that the receive signal in the RF path is rather week and sufficient amplification has to be provided, while the least amount of excess noise should be added. Hence, the amplifier of Fig. 12 has been reviewed and was deemed to be suited as RF input stage of the new topology. In Fig. 15, it can be identified in the dashed box on the left-hand side. Some adjustments - especially with respect to the values of components in the feedback network - had to be made, though, as the application required a shift in the covered frequency range. At that time, the prediction signal had to be provided by a current steering DAC and can be assumed to be of rather large signal swing. Hence, no amplification is provided for this signal. Promoted by the nature of the prediction signal, the required signal subtraction is performed in the current domain. In Fig. 15, two current mirrors inject their output signals into a common output node for this purpose. While this implies signal addition instead of signal subtraction at first glance, signal addition can be turned into signal subtraction by simple sign inversion, which is enabled either by exploiting the properties of the pseudo-differential amplifier structure itself or by sign selection of the prediction signal in the digital domain. The special current-mirror arrangement of Fig. 15 has been chosen for balancing the maximum output powers which is important to ensure that the signal from the DAC input can cancel the signal from the RF input. Linearity of both signal paths, transconductance from the DAC input to the common output and maximum output currents of the DAC have to be harmonized to account for this requirement.

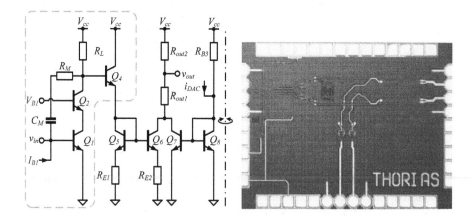

Figure 15. Half-circuit schematic diagram (left) and chip photograph (right) of the feedback-sampling LNA

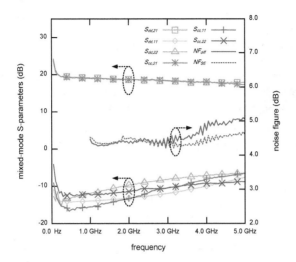

Figure 16. Measurement results for the RF path of the feedback-sampling amplifier. (The arrows point to the related axis scaling.)

This topology has been characterized in detail. Results for the RF signal path are presented in Fig. 16.

As explained for the pseudo-differential amplifier in section 3.2.3, the PM 8 probe station equipped with GSSG-probes can be used to examine an individual differential signal path. In order to detect the mixed-mode parameters given in Fig. 16, the device has been exposed to true mode excitation provided by the network analyzer, while calibration data have been applied to compensate for losses caused by the test set-up. Noise characterization can be performed by the method presented in section 3.3 and has been discussed in [19][20]. A remarkable feature of the results of this measurement is the fact that the differential noise figure NF_{diff} does not coincide with the noise figure measured from one signal path[6] NF_{SE}. Especially at higher frequencies, a large deviation occurs. In [20], this is explained by cross-talk caused by parasitics. So, the use of differential de-embedding schemes is recommended instead of a single-ended noise figure measurement from one signal branch. However, in view of the aim to assess the capabilities of the feedback-sampling approach, the signal subtraction itself is most interesting. In [27], we presented different test set-ups for verification purposes. Fig. 17 shows four representative results which confirm the ability to cancel the RF signal by an appropriate prediction signal.

---

[6] In Fig. 16, NF curve progressions start at 1 GHz because this is the lower corner frequency of the hybrid couplers used for measurement.

Those measurements were obtained from a test set-up incorporating the PM 8 probe station with GSSG-probes, in which two signal generators[7] synchronized by a frequency standard provided the input signals to both inputs at appropriate power levels via two hybrid couplers. The differential output signal of the DUT was recombined by a third hybrid coupler and displayed by a signal analyzer[8]. In Fig. 17, no loss compensation is applied and results are clipped to 100 kHz span. Two cases can be distinguished: First, the digital prediction signal has been switched off (DAC$_{off}$) and only the RF signal has been present at the inputs. Then, also the prediction signal has been applied (DAC$_{on}$) and a notable reduction in output signal power can be observed for all frequencies.

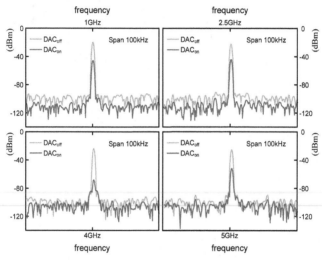

Figure 17. Signal subtraction enabled by the feedback-sampling amplifier.

3.4.2. Subtractor with Low Impedance antenna interface

The feedback-sampling amplifier of the preceding subsection is expected to perform well as long as the assumption of (reasonable) small input signals is justified. A key requirement in the feedback-sampling concept is linearity preceding the signal subtraction in order not to distort the zero crossings which are sampled by the analog-to-digital converter. However, as soon as array operation is considered, antenna cross-talk is likely to violate this assumption. In addition, a dense antenna array requires the antennas to have small outer dimensions. This can be achieved by dielectrically scaling the antennas, which - in turn - leads to a low (7 Ω) feeding point impedance. The latter has to be interfaced by the subtraction circuit. The topology shown in Fig. 18 is a first approach towards an analog subtractor which provides appropriate single-ended inputs to interface with both, a dielectrically scaled antenna and

[7] Rhode & Schwarz SMJ100A
[8] Rhode & Schwarz FSV

the DAC. In this implementation, noise figure is traded against linearity, as input signals close to 0 dBm might occur. For its implementation, component parameters have been determined by a semi-automated procedure, in which the input stage - a common base configuration - was optimized with respect to input power matching, while an upper bound for NF_{min} was respected and noise matching was clearly observed. As in the case of the feedback-sampling amplifier according to Fig. 15, signal subtraction is performed in the current domain using a common output node.

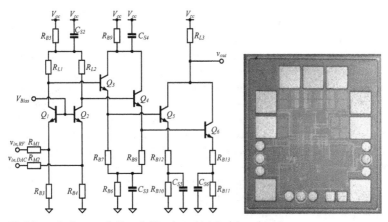

Figure 18. Schematic diagram (left) and chip photo (right) of the subtractor.

A printed circuit board has been designed to enable joint performance evaluation of this amplifier and the 7 Ω antenna. Due to the low feeding point impedance, separate characterization is less useful. To avoid problems involved in interconnecting devices with 7 Ω reference impedance, the amplifier should directly be attached to the antenna, which is supported by the board. Thus, both evaluation and refinement of this circuit will have to be performed in close collaboration with our partners from the *ultraMedis* project.

4. Transmitter circuits

4.1. Introduction

The circuits introduced in this section serve for the M-sequence topology. They have been implemented in a cost-efficient 0.25 µm Silicon Germanium BiCMOS technology, which opens up new fields of ultra-wideband radar applications. In the following sub-chapters, the design of different hardware blocks for the ultra-wideband radar front-end is presented. The design of a multi-purpose M-Sequence generator is presented which acts as a pulse compression modulator and exhibits an up-conversion mixer. A highly efficient power-distributed amplifier has been implemented utilizing a novel cascode power matching approach to achieve superior output power performance. Additionally, a fully differential broadband amplifier using cascaded emitter followers has been designed that exhibits a variable gain control and excellent broadband performance.

4.2. M-sequence generator

The well-known very broadband spectrum of M-sequences is widely used for testing the correct functionality of broadband integrated circuits, such as amplifiers, multiplexers, and transceivers. The run for higher data rates and amplifiers with broader bandwidth often outperforms commercially available test equipment and necessitates some sources to test these circuits. The measurement equipment vendors cannot supply data sources as fast as the technology evolves. The application which is targeted in this chapter is that M-sequences are used for pulse compression in ultra-wideband radar systems. For this application, it is important that the generator consumes little energy only, and it should generate a sequence of appropriate length (see (2)). Early high-speed PRBS generators for high data rates have been employed in III/V HBT technologies [28]. Moreover, a 110 Gb/s PRBS generator has been published in [29] using InP HBT technology with a transit frequency (f_T) more than 300 GHz. Recently, several PRBS generator circuits have been published in SiGe bipolar technology for test purposes in fiber-optic communications. In [30] a 100 Gb/s $2^7 - 1$ PRBS generator has been implemented in a 200 GHz f_T SiGe bipolar technology. As in the 80 Gb/s $2^{31} - 1$ pseudo random binary sequence generator introduced in [31], the output of the shift register has been multiplexed to achieve a higher maximum data rate. However, these circuits have a power consumption of 1.9-9.8W and utilize cost-intensive high-end processes. A 4×23 Gb/s $2^7 - 1$ PRBS generator with a power consumption of 60 mW per lane has been publicized in [32] utilizing a 150 GHz f_T SiGe BiCMOS technology. A $2^7 - 1$ multiplexed PRBS generator in 0.13 μm bulk CMOS exhibits 24 Gb/s output data rate [33]. In the following section, the circuit implementation with measurement results of the M-Sequence generator is presented.

4.2.1. Upconverted M-sequence generator

A simple way to generate M-sequences is to utilize a digital linear feedback shift register (LFSR), as depicted in Fig. 19. This device generates a binary pseudo-random code of length $2^n - 1$, where n is the number of stages in the shift register. Feedback is provided by adding the output of the shift register, modulo two, to the output of one of the previous stages. The actual sequence obtained depends on both the feedback connections and the initial loading of the register.

The proposed architecture depicted in Fig. 20 consists of serially connected shift registers with the characteristic polynomial

$$f(x) = x^9 \oplus x^5 \oplus 1 \qquad (23)$$

an additional XOR gate acting as a modulo-2 adder to yield the delayed sequence, and one multiplexer. The selected feedback in the proposed architecture enables to generate two M-

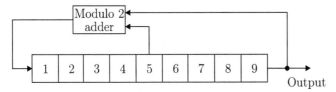

Figure 19. Nine bit linear feedback shift register for 9th order M-Sequence.

sequences with a mutual shift of half the word length. Those are multiplexed to yield the same M-sequence at twice the data rate. The input for the multiplexer is set between the two latches of the fifth flip-flop highlighted Fig. 20. This leads to a phase shift of half the pulse width in order to achieve the maximum voltage swing at the input of the multiplexer. Thus, the proposed architecture makes it possible to boost the circuit performance at the cost of an additional adder and a multiplexer. The architecture is extended to provide the possibility of upconversion for the generated M-sequence. This has been facilitated by implementing a mixer core at the output of the multiplexed LFSR. The mixer performs a BPSK modulation of the 9th order M-sequence signal generated by the multiplexed shift register. The circuit was implemented as an XOR gate instead of a conventional Gilbert cell as opposed to [34]. The actual circuit is nearly identical but the XOR operates in the limiting region compared to the small-signal operation of the Gilbert cell. The limiting behavior simplifies the design and requirements of the mixer, and results in lower power consumption. No emitter degeneration has to be implemented to increase linearity for large signal inputs. The XOR gate is driven by a LO buffer that can be digitally controlled to allow the generation of baseband M-sequence signals without the need for up-conversion. An additional output buffer with a resistively matched output has been included in order to control the output voltage swing in a wide range.

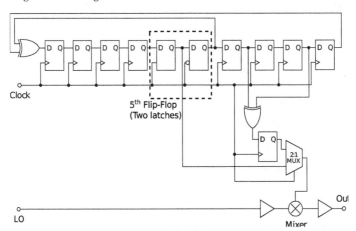

Figure 20. Circuit architecture of the proposed $2^9 - 1$ M-sequence generator providing the possibility of up-conversion for the radar signal.

Once the functional simulations have been completed, each individual block is designed on the transistor level. As the multiplexing architecture has been chosen, the flip-flops only have to work at half the operation speed. Standard CML flip-flops have been designed consisting of two latches, which inhibit two differential pairs. A schematic diagram of a CML flip-flop is depicted in Fig. 21. The flip-flops used in the LFSR are designed to offer a differential output voltage of 2×300 mV. According to [31], the tail current I_T and the emitter area A_e are related by

$$A_e = l_e \times w_e = \frac{I_T}{1.5 J_{peak\, f_T}} \qquad (24)$$

for fastest switching time such that $J_{peak\, f_T}$ is the current density for maximum f_T. The lowest tail current is set by the minimum allowed transistor size, which is $l_e \times w_e = 0.84\ \mu m \times 0.42\ \mu m$. Thus, the tail current is chosen to be 0.85 mA whereby the collector current for the maximum transit frequency f_T is 1.25 mA for a $84\ \mu m \times 0.42\ \mu m$ transistor. The output voltage swing of each flip-flop was set to 2×250 mV. Simulations indicate that the latches work up to 12.5 Gb/s, which is sufficient for a 25 Gb/s multiplexed M-sequence.

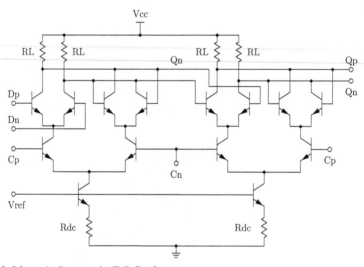

Figure 21. Schematic diagram of a CML flip-flop.

4.2.2. Simulated and measured results

The M-sequence generator has been simulated in time domain to find out the maximum data rate and to verify the correct function of the register. At clock frequencies higher than the maximum allowable clock frequency, the PRBS register does not work as expected and the output is not an M-sequence. A simulated waveform of a 25 Gb/s M-sequence is shown in Fig. 22 using the techniques described above. It can be seen that every bit can be distinguished from each other. A single bit has a slightly lower output voltage than a bit sequence with the same value, which is caused by the limited output bandwidth. As mentioned before, the mixer and the output buffer both have a limiting character which attenuates the flip-flop glitches. Thus, the waveform exhibits low ripple, actually at sequences with a series of equal bits, which indicates that the clock feed through is very low. However, the output waveform exhibits some deviations, which is quite common for circuits at this high data rate. This behavior may result from the slightly inductive behavior of an emitter follower in the signal chain. As long as the circuit is stable, this does not cause problems.

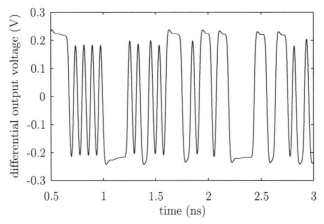

Figure 22. Simulated M-sequence waveform at 25 Gb/s data rate.

The M-sequence generator chip is placed on a Rogers TMM10i ceramic substrate for wire bonding. In order to protect the circuit mechanically and keep the bond wires as short as possible, it is placed in a topside cavity and fixed utilizing an electrical and thermal conductive epoxy glue, as shown in Fig. 23. The thickness of the ceramic substrate has been chosen to be 381 µm, which is almost equal to the chip height of 370 µm and the glue. Thus, the distance between the substrate edge and the bond pad can be reduced. The continuous ground plane below ensures a good thermal conduction, and 1.2 mm thick *FR*4 stabilize the brittle ceramic substrate.

The correct function of the generator can be checked through calculation of the normalized autocorrelation function of one complete M-sequence. The calculation of the ACF has been implemented in Matlab. As the correlation properties are of substantial interest for radar applications utilizing pulse compressed waveforms, the PRBS signal is measured with an Agilent DSO 91204A oscilloscope and compared with the simulation results. The simulated and measured 10 Gb/s waveforms together with the normalized cross correlation function of the measured signal are presented in Fig. 24.

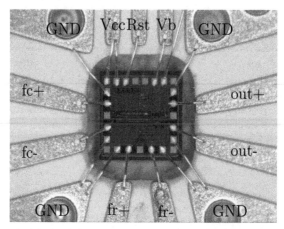

Figure 23. Photograph of the bonded M-sequence generator chip placed in a topside cavity on a ceramic substrate.

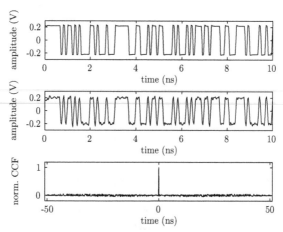

Figure 24. Simulated (top) and measured (middle) time domain representation of the proposed 10 Gb/s M-Sequence and their normalized cross correlation function.

4.3. Distributed power amplifier

The transmitted random sequence is subjected to high losses especially when transmitted through human body cells. Thus, a power amplifier is required to be placed directly before the antenna to increase the signal power. Distributed amplifiers (DAs) are appealing aspirants for UWB systems due to their inherently large bandwidth. The two major challenges in designing distributed power amplifiers are maintaining high linearity over the entire bandwidth, since narrowband linearization techniques cannot be utilized, and achieving high output power and efficiency. In order to increase the HBT distributed power amplifier performance, which is limited by the characteristics of the active cells used [35], alternative structures are investigated. The cascode cell is an appealing circuit due to its higher output impedance, higher breakdown voltage, and reduced Miller effect. Moreover, loading the two transistors by the required impedance for optimum power leads to an output power twice as high as compared to a single transistor. However, the conventional cascode configuration does not meet these conditions since the common base transistor's (T_{cb}) low-input impedance restricts the output voltage excursion of the common emitter transistor (T_{ce}). Therefore, it does not see its optimum power load impedance. In addition, the power performance of the cascode cell becomes one of the most important challenges to obtain maximum output power over the required bandwidth. To be power optimized, another series capacitor C_a is inserted on the base of T_{cb} to avoid its early power saturation compared to T_{ce}. A small signal model of the modified cascode gain cell is depicted in Fig. 25. The input impedance of the common base transistor can be calculated as follows:

$$z_{in,cb} = r_{be,2} + R_{b,2} + \frac{1}{j\omega C_a} \tag{25}$$

The capacitor C_a and input impedance $z_{in,cb}$ act as a voltage divider between the optimum values for $v_{ce,1}$ and $v_{be,2}$.

In order to achieve higher gain and greater bandwidth, an additional inductor is added between the collector of T_{ce} and the emitter of T_{cb}. The influence of a 1.5 nH inductor and various capacitances on the voltage gain of the cascode cell is demonstrated in Fig. 26. The small-signal schematic diagram of the modified cascode cell with inductive peaking is presented in Fig. 27a. The output resistance of the modified cascode circuit can be written as

$$z_{out} \approx z_{ce,2} + \left[\left(j\omega L_a + z_{ce,1} \right) | | \left(z_{be,2} + R_{b,2} + \frac{1}{j\omega C_a} \right) \right] \tag{26}$$

neglecting the influence of miller capacitance c_{cb}. This leads to a resonance effect which is dominated by L_a, C_a and c_{be}. The self-resonant frequency f_r of the LC low pass can be calculated as

$$f_r = \frac{1}{\pi \sqrt{\dfrac{L_a C_a c_{be,2}}{C_a + c_{be,2}}}} \tag{27}$$

The resonant frequency shows good agreement with the theoretical considerations set out in (27).

Another effect is that the output impedance of the cascode cell increases significantly from 2 to 15 GHz under the influence of the 1.5 nH inductor. The initial values for L_a and C_a are then optimized under large-signal conditions using nonlinear simulations of the inductively

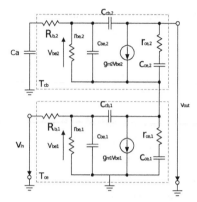

Figure 25. Small signal equivalent circuit of the modified cascode gain cell.

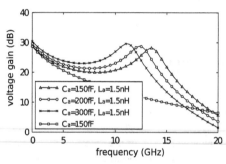

Figure 26. Voltage gain of different cascode cells with and without additional inductor $L_a = 1.5\ nH$ and various capacitances C_a.

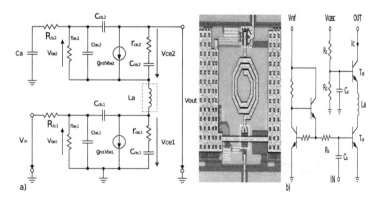

Figure 27. a) Small signal model of the modified cascode gain cell with inductive peaking and b) the schematic view.

peaked cascode circuit close to 1 dB compression point in order to obtain the maximum output power and efficiency. Accordingly, one single cell of a simple common emitter stage is used to synthesize the required ratio between the values of the inductor and capacitor. The test circuit is terminated with a 200 Ω resistor. The goal is to have equal deflections of the load lines both for the common emitter and the common base transistor. These results demonstrate that the proposed cascode configuration can obtain twice the output voltage swing compared to a single common emitter transistor at the same collector current so that twice the output power can be achieved.

A demonstrator chip has been implemented utilizing the methodology described previously. Fig. 28 depicts a schematic diagram of the four-stage tapered collector-line traveling wave amplifier with capacitive coupling and power-matched cascode gain cells. Each gain cell consists of two 26.4 μm^2 standard purpose transistors with BV_{ceo} of 4 V and a peak f_t value of 45 GHz, which are connected by a 1 nH inductor L_a. A single gain cell is depicted in Fig. 27b. The octagonal spiral inductor exhibits a Q-factor of 20 at 12 GHz.

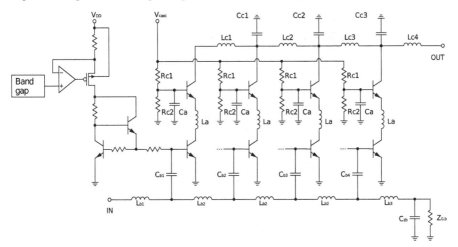

Figure 28. Complete schematic representation of the implemented distributed amplifier structure.

The tapered collector line has been realized using staggered inductors $L_{C1} > L_{C2} > L_{C3} > L_{C4}$ together with shunt capacitances $C_{ci} \| C_{ce}$; $(1 \leq i \leq 3)$ in order to achieve a coherent addition of the collector currents and a flat gain over the entire bandwidth. Biasing is implemented using three transistor current mirrors with ratio of 32:1 and a low dropout (LDO) voltage reference driven by a band-gap voltage source. A chip microphotograph of the complete circuit is shown in Fig. 29. The transistors are biased through the collector line by means of an external bias-tee. The bias point was selected at $VCC = 5\,V$ and $VDD = 2.6\,V$. A power and ground grid facilitates a low impedance connection and - due to the low distance between the congruent metal grids - a large capacitor is shaped. The chip size of the amplifier circuit is 2.1mm².

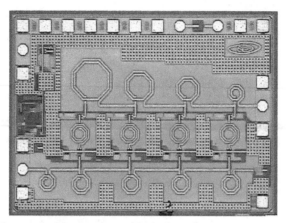

Figure 29. Microphotograph of the manufactured 2.1 mm² traveling wave power amplifier.

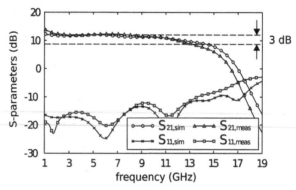

Figure 30. Simulated and measured small signal gain S_{21} and input reflection coefficient S_{11}.

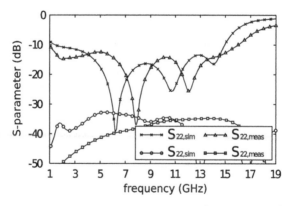

Figure 31. Simulated and measured output reflection coefficient S_{22} and isolation S_{12}.

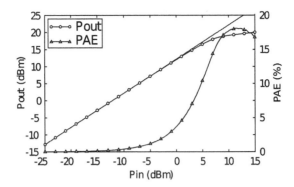

Figure 32. Output power P_{out} and power added efficiency PAE at a center frequency of 7 GHz.

The distributed power amplifier chip was tested via on-wafer probing. The measurements of the circuit were carried out using an Agilent N5242A PNA-X vector network analyzer. Fig. 30 shows the simulated and measured small signal gain S21 and input return loss S11. The traveling wave power amplifier exhibits a measured gain of 11 dB with a gain ripple of ±1 dB up to 12 GHz and a 3 dB bandwidth of 13 GHz. The simulated and measured output return loss S_{22} and the reverse isolation S_{12} are illustrated in Fig. 31. Both the input and output return loss are below -12 dB over the entire frequency range. The measured reverse isolation S12 remains below -35 dB. The circuit is unconditionally stable, also verified for large RF input signals. Fig. 32 shows the output power Pout and power-added efficiency PAE at a center frequency of 7 GHz. The 1 dB compression point is at 17.45 dBm with an associated power-added efficiency of 13.9%. The saturated output power P_{sat} is 20 dBm and the maximum power-added efficiency is 22.1%.

4.4. Differential broad band amplifier

Broadband variable gain amplifiers are key components for ultra-wideband radar applications and important building blocks to increase the dynamic range. Especially for M-sequence based radar systems without upconversion, the lower frequency range, which contains most of the signal energy [36], has to be considered. Biomedical and ground penetrating radar systems necessitate a lower frequency boundary of less than 1 GHz [37], [38].

Moreover, the broadband variable gain amplifier (VGA) should be fully differential. Great care has to be taken to avoid the distortion of the signal shape through gain ripple and group delay variation. In this section, the analysis, design and measurement results of a fully differential broadband VGA are presented. After some considerations about mismatching in broadband amplifiers have been made, the frequency behavior of cascaded emitter followers is investigated, and the implementation of a variable gain control is explained. Finally, the implementation of the broadband amplifier is presented, introducing the circuit architecture and presenting measurement results. The amplifier is

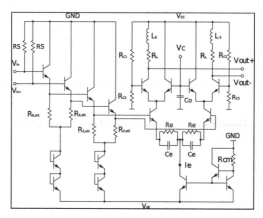

Figure 33. Simplified schematic diagram of the proposed broadband variable gain amplifier.

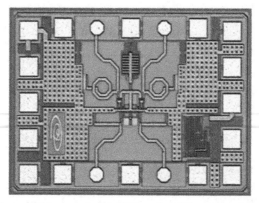

Figure 34. Microphotograph of the manufactured 0.75 mm² variable gain amplifier.

fully differential and based on a cascode configuration as depicted in Fig. 33. This is useful for high frequency circuit design, because this multi device configuration has small high-frequency feedback, achieved by the negligible Miller effect, and a large bandwidth. Driving the cascode stage with cascaded emitter-followers leads to an enhancement of bandwidth and provides dc level shifting [39].

The voltage gain of the cascaded emitter-followers has a frequency dependence that is similar to the frequency dependence of the transfer function of an RLC series resonance circuit [40]. This can be used to provide gain peaking at the desired frequency. The transfer function depends on the transistor parameters, the biasing current, the resistors, and the load. The main problem using emitter-followers to drive cascode stages is that the circuit might become unstable. This will be the case if the negative input resistance of the second emitter-follower stage becomes larger than the positive output resistance of the first stage at a certain frequency, which is shown in Fig. 35. The frequency is determined by the point where the imaginary parts, which are the reactances of the series resonator, cancel each other out:

$$Re\{Z_{11}(\omega_o)\} = -Re\{Z_{22}(\omega_o)\}$$ (28)

$$Im\{Z_{11}(\omega_o)\} = Im\{Z_{22}(\omega_o)\}$$ (29)

The core of the broadband amplifier is a signal summing VGA as illustrated in Fig. 33, where the gain is controlled by applying an analog dc voltage at V_C. The amplifier gain can be set from 0 to maximum gain whereby it behaves like a cascode differential stage when the control voltage is set to 0 and all current flows through the load resistors R_L. Furthermore, capacitive emitter degeneration is used to attain additional gain at high frequencies for a higher cut-off frequency. Tuning the values of R_e and $C_{e'}$, introduces trade-off between high gain, bandwidth and stability, because C_e influences the capacitive load of the cascaded emitter followers. Inductive peaking is carefully applied using small inductors L_C in the collector branches in order to avoid high group delay [41]. In order to achieve a high output swing, high currents in the differential amplifier are necessary.

The circuit is implemented in the 0.25 µm IHP SGB25V value technology. A chip photograph of the broadband variable gain amplifier is depicted in Fig. 34. The circuit elements composing the amplifier core have been arranged symmetrically to maximize the even mode suppression. The mixed-mode S-Parameters are measured on-wafer using 150 µm GSGSG probes.

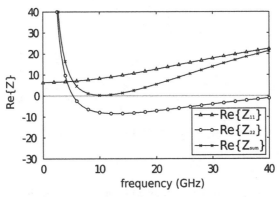

Figure 35. Simulated impedances Z_{11} and Z_{22} illustrating the stability relations of the emitter followers.

Fig. 36a illustrates the differential simulated and measured gain Sdd21 as well as the input and output return loss at 100 Ω differential source and load impedance. The measured differential gain is 11.5 dB with a gain flatness of ±1.5 dB. The 3 dB cut-off frequency is 30 GHz, which results in a gain-bandwidth product (GBP) of 113 GHz which is 1.5 times the

f_t of the transistor. The corresponding measured and simulated group delay is shown in Fig. 36b. The measured group delay variation is 35 ps, which is higher than that in the simulation and also induced by the stronger resonance behavior.

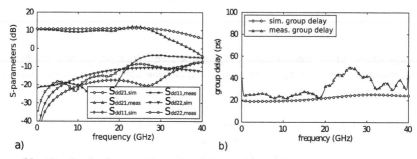

a)   b)

Figure 36. a) Simulated and measured mixed mode S-parameters. b) Simulated and measured group delay.

As depicted in Fig. 37, the amplifier gain can be adjusted between 0 and 11.5 dB. The large signal behavior is measured on-wafer. An output 1 dB compression point of 12 dBm has been measured up to 20 GHz.

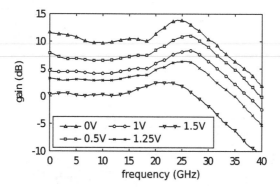

Figure 37. Measured gain by sweeping the control voltage V_c from 0 to 1.25 V

5. High-speed data capturing

5.1. Introduction of data capturing device using feedback principle

A straightforward data capture and digitizing can be directly performed by a conventional analog-to-digital converter (ADC). There are a number of limitations which arise using this method. The first and most crucial one is an inverse relationship between accuracy and speed of the conversion. In terms of the ADC, it is the inverse relationship between resolution and bandwidth. It is impossible to realize a high-speed ADC with the resolution which fulfils the sensor specification in modern technologies.

To overcome this limitation, a more complicated method of data capture based on "stroboscopic feedback loop" can be used. This method utilizes a feedback loop to relax accuracy requirements of the ADC (see [2] and chapter 6.2.3). The digital output of the data capturing device is represented by two summands: the value of the first summand is measured by the ADC; the value of the second summand is calculated based on its previous state and on the first one. The ratio between predicted and measured summands, i.e. between the resolution of the ADC and DAC can be calculated from the conversion efficiency of the both converters [42].

The block diagram of the data capturing device with feedback is depicted in Fig. 38. It consists of 3 logical parts, highlighted in colors in Fig. 38: Signal Processing, ADC and DAC, and LNA with subtraction amplifier. Although the subtraction amplifier belongs to the data capturing block, it has been integrated into LNA and moved to the receiver part of the sensor.

The data capturing device works as following:

- A capturing block digitizes a difference (residue) between the received and predicted values. This function is performed by a high-speed low-resolution analog-to-digital converter.
- A digital predictor evaluates the data from the ADC and makes a prognosis about the value to be expected next.
- The predicted value is converted into an analog form with a high-speed DAC.
- In analog domain, the predicted value is subtracted from the received signal with a subtraction amplifier.

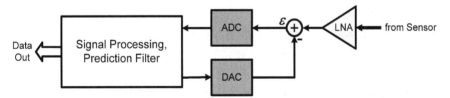

Figure 38. Block diagram of the data capturing device

5.2. High-speed 4-bit analog-to-digital converter

5.2.1. High-speed analog-to-digital converter

The fastest type of the A/D converters is a full flash ADC. A block diagram of a typical full flash converter is shown in Fig. 39. It consists of a reference network, a bank of comparators, correction and encoding logic and test buffers. The challenges of the implementation of the ADC are usually related to the analog part of the converter, namely to the reference network and to the bank of the comparators. It is possible to implement the high-speed comparator in the selected technology which will satisfy all requirements, but the reference network is a bottle-neck of the converter.

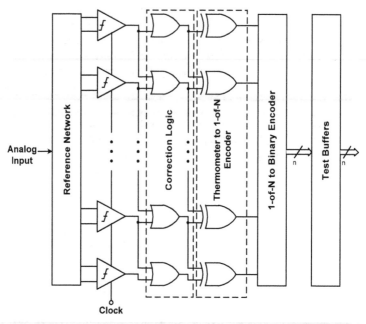

Analog→
Input

Clock

Figure 39. Block diagram of the full flash ADC.

5.2.2. Reference network

The task of the reference network is to provide equidistant reference voltages which will be further processed by the comparators.

There are two conventional implementations of the reference network. First, the simplest way is a Kelvin divider or resistor ladder. It suffers from several drawbacks, such as DC-bowing, clock and input feed-through [43]. Furthermore, is not well suited for the high-speed ADCs.

A second configuration is a differential one. It consists of two branches; each has a driver loaded with a chain of serially connected tap resistors. Both branches are equal, only outputs of the second branch are "inverted" or mirrored with respect to the middle point [43]. The main problem related to the differential network is its bandwidth, which often becomes a bottleneck of the system. The reference network has to drive a big parasitic capacitive load caused by the bank of comparators. In the full flash ADC, it is one of the main limitations, because the number of comparators is doubled when increasing the resolution by 1 bit.

The second problem of such network is the non-equal transfer characteristic of the output nodes [42].

5.2.3. Proposed bandwidth enhancement technique

Drawbacks of the conventional differential reference network are mainly due to its serial configuration; a change in one component will affect the others. This inherent property of the serial connection makes individual adjustments and compensations impossible. To overcome this limitation, a new configuration of the reference network is proposed. An idea is to build the resistor network in a segmented serial-parallel configuration and substitute one driver (emitter follower) with several drivers, connected in parallel. A full overview of the possible configurations is described in [44]. Among this variety one configuration should be highlighted, namely the configuration illustrated in Fig. 40 where each segment contains one tap resistor and one current source. The reference network is fully parallel, thus allowing the maximum speed to be achieved.

The main feature of the parallel network is the flexibility to choose component values. This freedom gives the possibility of equalizing the bandwidth of an individual segment that leads to the optimal speed at given power dissipation.

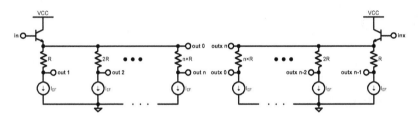

Figure 40. Parallel configuration of the network with one resistor per segment.

Fig. 40 shows the case when all driver currents are equal. In practice, it is more useful not to keep the currents in all segments equal, but to equalize the bandwidths in each segment instead. The network, however, does not only present good advantages, it also has some drawbacks. Flexibility of adjusting different parameters leads to different geometries of the resistors. In the case of the conventional network, all resistors have the same value and the same geometry. Proper layout minimizes the mismatch between them. The proposed network cannot benefit from this feature.

5.2.4. Design of comparator

Signals from the reference network are led to a bank of n comparators. Comparators decide if the input is above or below the reference. For decreasing the probability of errors, a master-slave comparator with a preamplifier is used. An overall schematic diagram is shown in Fig. 41. The role of the preamplifier for the comparator is twofold: It works as a limiting amplifier, and it provides an additional amplification of the input signal. Another important function of the preamplifier is isolating the reference network from kick-back noise, produced by the master latch. In this particular example, the Cherry-Hooper amplifier with emitter follower feedback is used as preamplifier.

The master latch has an auxiliary current source I_{aux}. This current source prevents the cross-coupled differential pair from being completely switched off, thus keeping the base-emitter capacitance charged. The time to charge this capacitance is decreased, and as a result the overall speed of the latch is increased. The I_{aux} has to be sufficiently small because it adds hysteresis which decreases the sensitivity of the comparator. Setting the value of I_{aux} equal to 10 % of I_{EE2} is a good compromise between speed and sensitivity. In the slave latch, there is no auxiliary current source because the input signal of the slave latch is relatively large, and an auxiliary current source does not have a strong influence as in the case of the master latch.

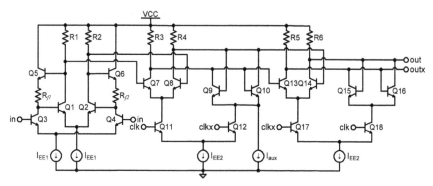

Figure 41. Master-slave comparator, full schematic diagram.

5.2.5. Experimental results

The ADC with the proposed parallel reference network was implemented in 0.25 μm SiGe BiCMOS technology. The Chip micrograph of the ADC is depicted in Fig. 42.

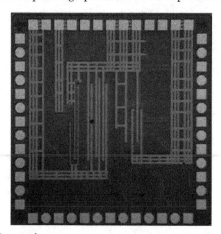

Figure 42. ADC's chip micrograph.

Static measurements: For measuring static errors of the ADCs, a low frequency 50 MHz sine signal was applied to the input of the converter at 5 GS/s sample rate. A deviation of a transition from the mean value, the differential nonlinearity (DNL), was calculated for each step. A cumulative sum of differential errors represents the integral nonlinearity (INL). The results are graphically presented in Fig. 43.

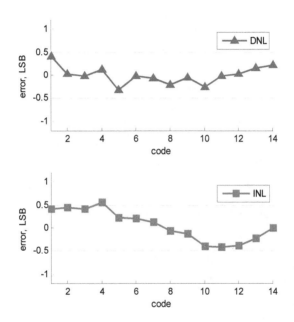

Figure 43. DNL and INL from histogram testing of ADC.

Dynamic measurements: Signal-to-noise and distortion ratio (SINAD) of the test circuit was measured over the frequency range up to 6 GHz at a constant sample rate of 15.01 GS/s. The small frequency offset of 10 MHz was made to accumulate quantization errors over the whole dynamic range. The measurement results are presented in Fig. 44, which shows SINAD of the converter up to the input frequency of 6 GHz. The dashed line shows a level where SINAD drops 3 dB below its value at low frequency. The frequency where SINAD crosses the 3 dB line indicates the effective resolution bandwidth of the converter, which in this case is greater than 6 GHz.

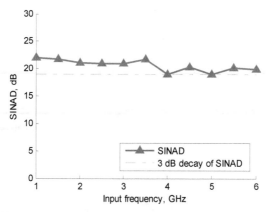

Figure 44. Measured SINAD over a frequency range up to 6 GHz.

5.3. High-speed predictor

The main function of the predictor of predicting the part of the received value was described above. The predictor also carries out two additional functions:

- Making subsequent averaging of the digitized values, increasing signal-to-noise ratio of the measured signal.
- Decreasing the data throughput for further data processing.

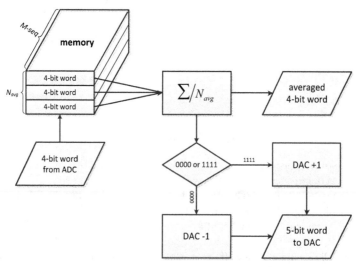

Figure 45. Functional diagram of the predictor.

The functional diagram of the predictor is depicted in Fig. 45. The predictor consists of the memory where the data from the ADC are accumulated; an averaging block, which makes averaging of the accumulated data; and a block where the output DAC value is calculated. An algorithm to calculate the DAC value is a modified version of the successive approximation algorithm with a constant ±LSB step.

The described functionality is coded using VHDL language and implemented using ECL library available in IHP BiCMOS Technology. For speed purposes, the predictor was divided into several sub-blocks which were implemented separately. This method decreases the complexity of the separate sub-block, and achieves a higher operational speed. The block diagram of the predictor is depicted in Fig. 46. The predictor consists of a demultiplexer, a bank of predicting blocks and a multiplexer. A predicting block carries out three functions: accumulation, averaging, and prediction. The demultiplexer deserializes the M-Sequence and commutates M-Sequence parts (chips) to the individual predicting blocks so that each has to work with only one defined chip. The multiplexer reverses the parallel processing and serializes the predicted values which finally fed the DAC.

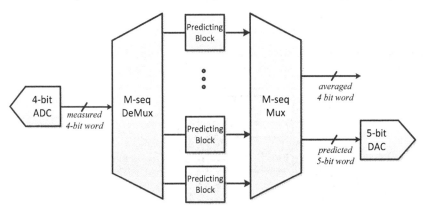

Figure 46. Structure diagram of the predictor.

5.4. High-speed digital-to-analog converter with off-chip calibration

The digital-to-analog converter transforms a digitally predicted value into the analog domain. To prevent information loss, the accuracy of the DAC should correspond to the accuracy of the whole capturing device. Simultaneously, the DAC should work at 10 GS/s. To satisfy both requirements, the converter is implemented using a segmented current steering architecture. The block diagram of the converter is depicted in Fig. 47. It consists of the two segments: A unary sub-converter and an R-2R sub-converter. The current sources of the both sub-converters are connected to the summing node. As will be seen later from measurements relying only on technology, component matching would give insufficient accuracy, which in this particular case is 10 times lower than required. Therefore, an additional calibration of the current sources is implemented. The current sources are realized as voltage controlled current sources. The controlling voltages are produced by auxiliary low-power µDACs which are externally controlled via SPI interface.

The calibration algorithm could be characterized as successive approximation of the DAC output to the reference value. The detailed calibration flow of the each current source is as follows:

1. The current source under calibration (CSUC) is disconnected from the summing node. For this purpose, the corresponding digital input is applied to the DAC.
2. The analog output of the DAC is measured and stored in memory as "zero-value". The measurement is performed with a 14-bit ADC on an FPGA board.
3. The CSUC is connected to the summing node.
4. According to a binary search algorithm, the MSB of µDAC is set to "1".
5. The output of ADC is measured again, and the difference between the stored "zero-value" and the measured value is calculated.
6. Depending on this difference, the decision concerning the value of the MSB of the µDAC is made.
7. Steps 4-6 are repeated for the remaining 9 bits of µDAC.
8. Steps 1-7 are repeated for each current source.

A set-up to implement the proposed calibration scheme is depicted in Fig. 47. The calibration algorithm is implemented on a Spartan-3AN Starter Kit board.

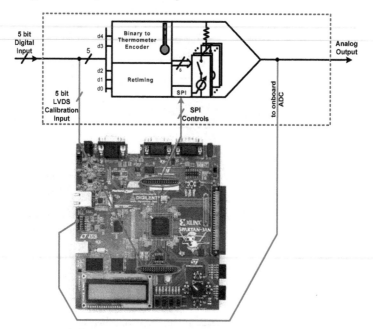

Figure 47. Off-chip calibration of the DAC.

5.4.1. Experimental results

Static measurements: The DAC test chip was implemented in SiGe 0.25 μm BiCMOS IHP technology. The chip was mounted on a test board and connected to the FPGA board. The results of static tests are given in Fig. 48 where both DNL and INL values before and after calibration are given. INL errors were also recalculated in percent of the input range. To

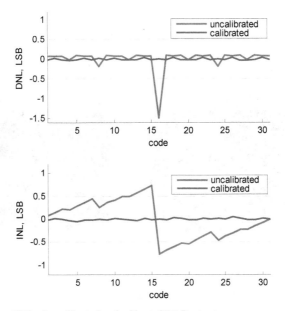

Figure 48. DNL and INL of uncalibrated and calibrated DAC output.

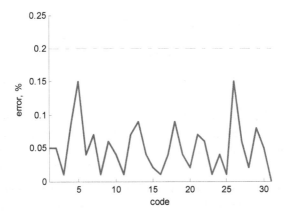

Figure 49. Static accuracy of the calibrated DAC.

achieve 9-bit overall resolution, the DAC should have an error below 0.2 %. The measured error after calibration is depicted in Fig. 49. The error is below 0.15 %.

Dynamic measurements: Dynamic characteristics of the DAC were measured together with the 4-bit ADC under the assumption that LSB usually works faster than MSB. The direct measurement of the spurious free dynamic range (SFDR) has no practical sense since the ADC limits the overall performance. For estimating the performance, an envelope test was applied [45]. Proper work of the converters assumes the presence of the all transition steps at the frequency of interest. Fig. 50 shows DAC outputs at 5 GHz and 5.5 GHz. Both converters (ADC and DAC) have all 16 transition levels up to 5.5 GHz input. Only the amplitude at 5.5 GHz starts to decay.

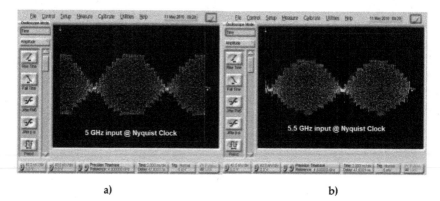

a)   b)

Figure 50. Envelope test of ADC-DAC at **a)** 5 GHz and **b)** 5.5 GHz.

5.5. Conclusion

The design and measurements of the high-speed data capturing device for the M-sequence sensor are described in this chapter. The data capturing device utilizes the "stroboscopic feedback loop" for achieving high dynamic range together with high sampling rate.

A number of different techniques are used to achieve the desired performance of the separate components.

To achieve a high effective resolution bandwidth of the analog-to-digital converter, the new segmented reference network was proposed. The new network, implemented in the ADC [46] allows increasing the effective resolution bandwidth several times compared to the similar conventional one [47], while the power dissipation is only slightly increased.

The high-speed predictor was described in VHDL and implemented using a high-speed ECL library. Despite the disadvantage of the power dissipation, the ECL implementation allows speeds of up to 10 GS/s to be achieved. Furthermore, it is simple to modify the predictor to comply with different system parameters, such as the M-sequence length or averaging factor.

An off-chip calibration was implemented for the high-speed digital-to-analog converter. The calibration is implemented on an FPGA-board. After having been modified slightly, it could be integrated into the DAC. The static errors of the DAC after calibration are lower than 0.15 % which allows the use of a converter in the data capturing device with a target resolution of 9 bits.

6. M-sequence devices

6.1. Introduction

While previous sections were aimed to discuss specific sub-components such as individual semi-conductor chips of an UWB-sensor, we would like to consider some aspects of the whole sensor electronics here. For that purpose, several M-Sequence devices were implemented at different integration levels, and some Ukolos-partners (*ultraMedis, CoLoR*) were provided with demonstrator devices for their own use. In order to have a running sensor system, the device implementation has to cover the whole manufacturing cycle from chip-design and manufacture, chip housing, RF-PCB-design and assembly, design and implementation of the digital components (ADC, FPGA, interfaces etc) up to the programming of sensor internal pre-processing, the data transfer to the host PC and application-specific software for data evaluation and visualization. Furthermore, device specific test and evaluation methods and routines had to be developed and implemented in order to perform high-resolution device characterization (e.g. [48])

In what follows, we will first introduce an experimental device which is aimed to evaluate new concepts or modifications under real conditions. Secondly, we refer to a device configuration which implements the principle depicted in Fig. 4 for the practical use by other Ukolos-projects and finally, there will be some discussions toward single chip solutions.

6.2. Experimental demonstrator device

6.2.1. Device concept and aim

The aim of an experimental demonstrator device is to investigate the impact of individual sub-components on the performance of the whole device, as well as to have the opportunity to flexibly perform device modifications without the need of redesigning complex RF-PCBs. The device is organized in a modular concept as symbolically depicted in Fig. 51. Fig. 52 shows an example of a demonstrator implementation of such kind.

The individual sub-components as e.g. shift register for stimulus generation, T&H-circuits, RF-power distribution, RF-synchronization etc. are organized as plug-ins. Hence, one can simply replace a device component by a new one if improved circuits, better IC-housing or

RF-PCBs are available. Furthermore, the various modules may be interconnected to different device structures as shown in Fig. 4 or Fig. 55.

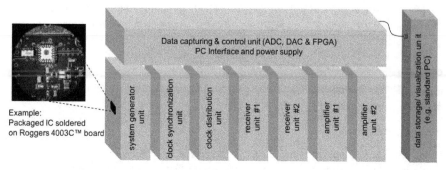

Figure 51. Modular concept of the experimental system.

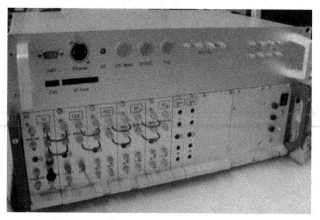

Figure 52. Photograph of the demonstrator implementation example.

6.2.2. Demonstrator performance

The particular RF plug-ins of the demonstrator are designed to operate with signals of large fractional bandwidth at the lower end of microwave frequencies or with toggle frequencies up to about 20 GHz. The generator unit provides periodic M-sequences of length 2^m-1, where m represents the order of the sequence. The demonstrator has optionally implemented 9th or 12th order generators, which accordingly produce signals with periods 511 or 4095 chips. The generator plug-in operates with toggle rates between 500 MHz and 20 GHz for the 9th order M-sequence, and the 12th order device may be operated between 500 MHz and 16 GHz. In the case of radar applications, the unambiguity range (4) of the measurement may cover values from 3.8 m (related to 9th order M-sequence and 20 GHz clock) up to 1.2 km (12th order M-sequence and 500 MHz clock).

The clock synchronization unit which precisely defines the receiver sampling points is a 9th order binary divider with a maximal toggle rate of 24 GHz. Random fluctuations of the sampling point (jitter) could be reduced down to some tens of femtoseconds [48] due to the balanced circuit topology and the optimized architecture of the timing system (see [2], [49] detailed discussions). Note, that the time position uncertainty of the measured impulse response (compare Fig. 5) is father decreased as consequence of the impulse compression (i.e. correlation; see [2] for discussion).

The clock distribution plug-in is an active device which recovers and distributes the sampling clock among the receivers and the analog-to-digital converters. The unit can handle clock pulses with 20 ps falling/rising edges and features wideband reverse signal rejection better than 40 dB per branch.

The receivers are ultra wideband sampling gates with an 18 GHz analog input bandwidth, better than -40 dB signal feed-through over the full bandwidth, -15 dBm input compression points and a decay rate of about 20 % per ms relative to full scale (i.e. 5...200 ppm per sampling cycle depending on the clock rate (0.5 – 20 GHz) of the system). Other potential components of the experimental demonstrator are discussed in sub-chapters 3 to 5.

The transmitter-receiver and receiver-receiver cross-talk is better than 130 dB over the full operational band. In order to achieve this value, attention was paid to RF-housing, clock signal distribution and power supply decoupling (see also Fig. 57). The recent configuration of the demonstrator RF electronics is able to handle (internally) up to about 70 000 IRFs per second (9th order M-sequence at 18 GHz system clock). The data transfer to a host PC (based on commercial standard interfaces like USB and LAN) reduces, however, the actually achievable update rate to about 300 IRFs/s. The corresponding gap is filled by synchronous averaging in order to use the available data amount for noise suppression. The achievable receiver dynamic is about 114 dB @ 1 IRF/s. It has to be noted that device non-linearity is classically qualified by the intercept point which is based on a Taylor-series model of the device under test and sine wave stimulation. In order to keep this established philosophy, the approach was extended to wideband signals [48].

This is illustrated by Fig. 53. In the example at the top, the Tx- and RX-port of an M-sequence device were connected via a variable attenuator and the impulse response was recorded for attenuator values between 0 and 120 dB. In the case of very weak input signals (large attenuation), we can only observe noise and device internal cross-talk. If we reduce the attenuation, the wanted signal peak (it is called "main pulse" in Fig. 53) appears and increases linearly with the signal level while the cross-talk level remains constant. By reducing the attenuation further, other signal parts become to protrude from noise. They also increase linearity at the beginning. These signals are caused from device internal reflections, deviations from the ideal time shape of an M-sequence and misalignments of the ADC timing (refer to Fig. 54). We call them device internal clutter. For very high signal levels, the receiver will tend to saturate which leads to the compression of the main peak and the internal clutter signals. Furthermore, the appearing non-linear distortions create new signal peaks which leave an apparently chaotic mark (see [2] and Fig. 1 for details). While the cross-talk and the internal clutter may be removed by device calibration [50] since they are caused by linear effects, the non-linear distortions should be avoided by respecting corresponding input levels of the measurement signal.

The level diagram at the bottom of Fig. 53 refers to the non-compressed receiver signal. It shows the strength of the linear, quadratic and cubic signal parts in dependency from the signal power (see [48] for details).

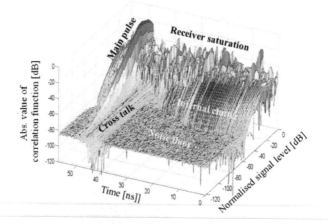

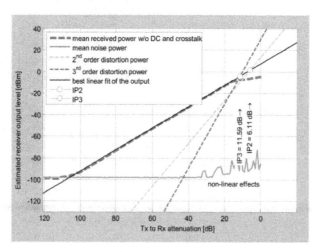

Figure 53. Top: Sensor pulse response of system as a function of the input level. Bottom: Level diagram (see also Fig. 1) of raw data (i.e. without impulse compression). In the shown case, the input related 1dB compression point is -14 dB below the transmitter power.

The effect of ADC timing misalignment is illustrated in Fig. 54. Theoretically, the ADC could capture the voltage sample at any time point within the hold phase of the T&H-circuit since by definition the signal level should keep a constant value during the hold interval. Unfortunately this is not case as demonstrated by Fig. 54. Here, the impulse response (i.e. the correlation function) of the M-sequence device was recorded by insertion a variable delay between the start of the hold phase and the trigger of the ADC. Ideally, we should see only a single pulse as long as the ADC is triggered within the hold phase and noise within the track phase (which is however out of interest here). But actually, some spurious signals appear whose strength and time position depend on the delay between "hold-start" and ADC-trigger. Hence, by selecting a reasonable delay between T&H and ADC, we can minimize these spurious signals.

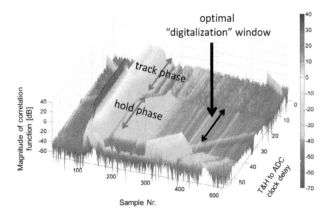

Figure 54. Sensor pulse response as a function of delay between T&H and ADC.

6.2.3. M-sequence feedback-sampling

This sub-chapter gives an example of the usefulness of the modular experimental device. It deals with feedback sampling. Feedback loops have been used for a long time in sampling circuits. However, they were usually restricted to sequential sampling having very large Nyquist rates so that only minor signal variations between consecutive samples appear. Only these variations are captured by that approach (see [2] for details).

In our case, this simple method cannot be applied since the voltage steps between two consecutive samples may cover the full receiver input range as we firstly apply Nyquist sampling and secondly, the natural order of the data samples may be disrupted due to interleaved sampling. Hence, we need some modifications of the principle which pose some challenges to the practical implementation.

For the purpose of feedback sampling, the data capturing & control unit was additionally equipped with a digital-to-analog converter which has to provide the feedback signal. The principle and the device structure are depicted in Fig. 55. The idea behind the digital feedback sampling implementation is to deal with high-speed signals (analog and digital) of low dynamic range (i.e. low amplitude) and to exploit the fact that the temporal variations of the scenarios under test are of the orders smaller than the measurement speed. This implies for the radargram (see Fig. 55, on the left) that adjacent samples at a horizontal line undergo only minor variations (instead of consecutive samples in sequential sampling). Thus, it will be possible to predict the measurement values along the observation time axis. This is the reason to insert a DAC into the feedback loop which converts the predicted digital values into analog ones. If the predicted signal levels are subtracted from the received signal, only the prediction error has to be captured by the ADC and processed by a digital high-speed system.

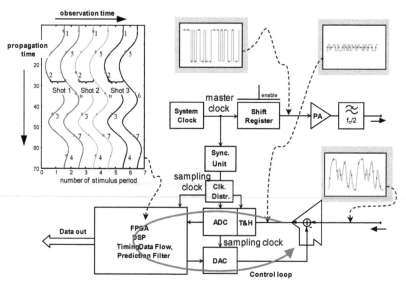

Figure 55. Basic structure of an M-Sequence feedback approach.

Fig. 56 gives an example of the output signals of the T&H-circuit. The constant voltage during the hold phase must be captured by the ADC. In the open loop example (above), we can observe that the hold voltage jumps from sample to sample. Hence, the ADC must be able to convert voltages within a large range. The second example shows the closed loop operation. Now, the predicted value is subtracted before AD-conversion, and we actually get a voltage during the hold phase which is always at about the same level. Under optimum conditions, the magnitude of the prediction error is determined by the strength of random noise which is usually quit weak. Therefore the requirements onto the dynamic range of the receive electronics can be relaxed.

Under optimum conditions, the magnitude of the prediction error is in the same order as random noise. Therefore, the demands made on the dynamic range of the receive electronics can be relaxed.

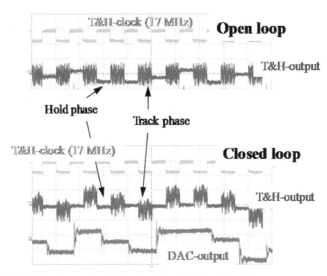

Figure 56. Digital feedback sampling. Above: open loop; below: closed loop.

6.3. Prototype devices

Fig. 57 shows a photograph of a primary (1Tx 2Rx) M-sequence RF board and corresponding ADC PCB with PC Interface (USB). The RF board is designed for assembly with the *HaLoS*-project originating ICs. Each of the board layouts corresponds to the architecture shown in Fig. 4, so that both boards connected together represent the basic M-sequence working unit. This unit is considered as main integral part of the UWB devices provided for partner projects within the UKoLoS- and other scientific projects.

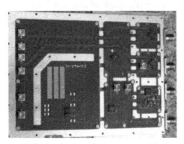 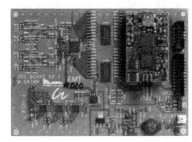

Figure 57. Photograph of the primary 1Tx 2Rx RF board (left-hand side) and corresponding ADC board (right-hand side).

While in the early project phase the sensor devices were finalized in cooperation with Meodat GmbH (Ilmenau, Germany), the final device assembly was performed by ILMSENS (TU Ilmenau service GmbH, Ilmenau, Germany) later. To give the reader an impression, some device examples are depicted in Fig. 58 as well in the chapter 11.

Figure 58. Left-hand side: 1Tx 2 Rx UWB sensor unit (MEODAT). Middle and right-hand side: 1Tx 2Rx UWB sensor unit with portable power supply and UWB reflectometer (ILMSENS).

6.4. Single-chip sensor head

The ability to create an optimized multi chip sensor is apparent, but the manufacturability of such system is much more difficult with a longer parts list and more complex assembly as for instance in the case of the construction of complex MIMO sensing systems (see 8Tx 16Rx system in the chapter 11). One promising way is to realize all active high-frequency system components (i.e. components on the primary RF board – see Fig. 57 left-hand side) onto one chip. This will enhance the overall system performance, reliability, robustness and assembling yields. By contrast, however, increased complexity on the single die means more second-order effects that have not been studied so far. For example, undesired on-chip coupling interactions between the different constituent system components become more pronounced and are more challenging to manage especially because of dealing with ultra-wideband signals. Such unwanted signal coupling or cross-talk can degrade the performance of the sensitive receive circuitry and, consequently, of the whole system. The aim to study such interactions which have not been considered so far, the expected advantages but also the knowledge gained from multi-chip approach analyses, have motivated the first monolithic integration of the complete RF-part of the M-sequence UWB radar electronics into one silicon die.

6.4.1. UWB single-chip head architecture

Fig. 59 shows the simplified block topology of the realized M-Sequence based single-chip transceiver head (alias System-on-Chip, SoC head). In correspondence with the system topology depicted in Fig. 4, the M-sequence transceiver SoC contains one transmitter and two receiver circuits (commonly assigned as 1Tx 2Rx configuration).

According to our experience, the 1Tx-2Rx topology of the primary sensor cell represents the optimum regarding achievable performance and circuit complexity. Moreover, the implementation of 1Tx 2Rx structure on one die has the advantage of permitting both cross-talk investigations between active and passive circuit parts (i.e. transmitter and receiver) as well as between two passive parts (i.e. receiver 1 and 2). From a practical point of view, the stand-alone 1Tx 2Tx devices are suitable for implementations where two receive channels are needed a priori, e.g. for simple localization tasks or in material testing (see chapter 11) in which the second (slave) receive channel can be used for device online calibration purposes. The desired MIMO usability as for instance in novel UWB-arrays for high-resolution near-field imaging (*ultraMedis*) or localizations (*CoLoR*) with 1Tx 2Rx constellation of primary sensing cells is also given.

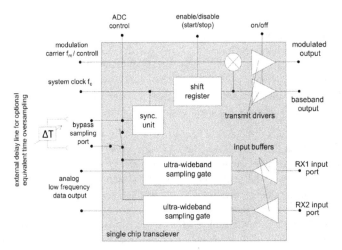

Figure 59. Simplified architecture of a fully monolithically integrated UWB M-Sequence based transceiver.

6.4.2. Design philosophy

It is apparent that the presented single chip architecture envisages ultra-fast switching cells (i.e. stimulus generator and synchronization unit) with their relatively high signal swing output buffers as well as very sensitive analog input blocks integrated on the same chip substrate. So, the undesired signal coupling or cross-talk can degrade the performance of the sensitive receive circuitry and, consequently, of the whole system. Especially in the case of analog devices, which handle the ultra-wideband signals, the on-chip interferences can be catastrophic. For example, intermodulation/interaction of noise components with the measured signal within the frequency band of interest may cause device saturation. Therefore, special emphasis is put on the isolation of the SoC channels during the design phase ,as discussed in [2] or [49].

6.4.3. Individual functional block peculiarities

Particular functional circuit cores of the SoC transceiver components are designed to fulfill at minimum the parameters of the demonstrator plug-in blocks discussed above. Additionally, the SoC transceiver includes additional build-in options to open further functionalities as e.g. (equivalent time) oversampling [51] or frequency conversion in order to meet the UWB radiation rules [52]-[55]. In particular, the SoC concept is intended for very wideband material investigations and MIMO-applications like in medical microwave imaging [56]-[61].

In summary, the goals of the single-chip integration are:

- to improve the synchronization between transmitter and receiver due to shorter interconnections with steeper signal edges,
- to provide means of a flexible adaptation of the operational frequency band by introducing a wideband modulator,
- to save power consumption by avoiding power hungry PCB-interconnection lines,
- to investigate broadband signal leakage on chip and cross-talk due to the housing,
- to avoid temperature effects on calibrated sensor systems due to temperature difference between the measurement channels and temperature expansion of device internal cables.

Thus, to achieve the desired MIMO usability, the shift register may be enabled and disabled, and transmitter buffers can be switched off (power down) by simple TTL-signals so that no external RF-switch is required to operate in a MIMO system. The transceivers are designed in such a way that they may either work while being driven individually or they may be cascaded with respect to the master system clock so that all units of a MIMO array work synchronously. Once the array is calibrated, a power down feature will be used for active transmitter selection. Thus, all receivers of MIMO array work in parallel and capture permanently data in order to get maximum measurement speed. As shown in Fig. 59, the transceiver IC is equipped with a wideband multiplier which optionally allows the sensor stimulus frequency band to be shifted and doubled ([2] or [49], [55]) or the operational band to be adapted to a specific application [50] in conformity with regulation requirements [52] - [54]. The channel is designed for operation up to 18 GHz.

Moreover, the multiplier can invert the stimulus M-sequence by implementing simple ECL signals on the control port. This feature may be useful to provide uncorrelated transmit signals in MIMO arrays. In addition, the sampling timing control chain is equipped with optional switchable shunt path. This add-on allows direct clock supplying from chip periphery. Thereby, user-selectable sampling rates or enhanced signal capturing approaches (e. g. equivalent time oversampling approach [50]) are possible without IC redesign. The analog receivers are designed to operate with wideband signals up to 18 GHz. The maximum linear operation input signal peak-peak swing is 60 mV.

Fig. 60 shows the chip die micrograph of the discussed transceiver with marked particular functional blocks and well visible top metal of a decoupling guard between the transmitter and receiver (line in the middle). The transmitter and receiver cores as well as their particular I/O pads are placed on the opposite die sides to minimize mutual on-chip coupling as well as inductive coupling between the bond wires after packaging. As extensively discussed in [62] or [63], [64], the decoupling guard is a guard well in a trench between the noisy transmitter and sensitive receivers. In the final assembly, the guard is connected to the quiet potential in order to fix the voltage of the substrate between the Tx and Rx die part by absorbing potential substrate fluctuations. The transceiver die occupies an area of about $2000\mu m \times 1200 \mu m$ and the build-in circuits consume in total about 300 mA from 3 V supply.

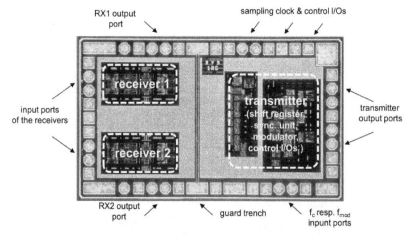

Figure 60. Microphotograph of the SiGe M-Sequence based transceiver head die with depicted particular functional blocks. The die size is 2×1.2 mm².

6.4.4. Single chip transceiver head evaluation

For the sensor head prototype evaluation, the transceiver chip has been measured on wafer as well as packaged with well-established chip-on-board technology using an optimized composite 4-layer carrier made from Rogers 4003C™, FR4 laminate and ultrasonic bonding procedure with 1-mil aluminum wires. The bond wire landing areas for RF ports on the board are designed to match as closely as possible (i.e. realizable) the pitch on the IC to avoid long wire connections. The cavity approach has not been implemented because of challenging technological realization on the selected carrier board. Fig. 61 shows the test board whereat the wired die is zoomed out for better visualization. The die is located in the center of the photo and top glue is used to protect the bond wires. It is mounted on a metal patch which is connected with VEE. This allows a direct connection between the substrate and the board's lowest potential. The top layer is mainly used for RF signal routing whereas the bottom layer is used for control lines. The inner plane layer below the top layer is the common board GND, and it also provides the GND reference for the RF signal lines. The other inner plane is the supply layer. The supply is bypassed to the GND with a 0.1 µF ceramic capacitor placed as close as possible to the die.

A basic test set-up for the sensor head prototype parameter evaluation is symbolically depicted in Fig. 62. The photograph on the right side shows an example of such test assembly. The evaluation board is connected to a 10-bit data "digitizer" (ADC), FPGA control and pre-processing unit which is equipped also with the PC interface. Moreover, for the parameter characterization, stable sinusoidal reference has to be connected to the system clock port (not shown in the photograph). This signal comes on board through an SMP connector and toggles the analyzed assembly. The toggle rate for the packaged prototype can be chosen quite flexibly between 0.5 and about 19 GHz, which implies a good compliance with the actual and intended applications needs.

Figure 61. Evaluation board for Single-Chip sensor head. The wired die (close-up photograph) is protected by top glue.

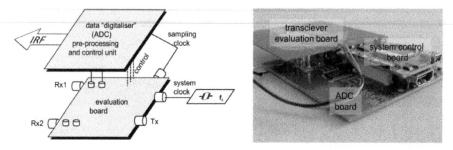

Figure 62. Generic transceiver test configuration (left) and an example of an experimental M-sequence based sensing unit assembly (right).

Interchannel cross-talk plays an important role in many applications. As Tx-Rx-decoupling up to 130 dB could be reached if the individual components are properly shielded (see Fig. 57, left), an interesting question is how the single chip devices behave with respect to that problem even though decoupling design techniques are implemented [2]. Fig. 63 shows the results for on-wafer measurements and the housed chip. Obviously, the chip design outperforms the quality of the chip wiring with respect to the cross-talk performance. The impulse response function of the housed chip is also shown in Fig. 63 (left). It was gained using the configuration as depicted in Fig. 62 (left). The cross-talk pulse can clearly be identified. However, it should be noted that it can largely be suppressed by post-processing via system calibration. Thus, achieved spurious free system dynamic range is comparable to that of the demonstrator device.

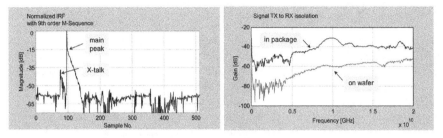

Figure 63. Example of normalized IRF captured with a 9th order M-Sequence based experimental single chip assembly and broadband measured TX to RX isolation.

As a result, it can be concluded that we have successfully realized a novel functional hardware platform, both multi-chip and single-chip based, for current (e.g. [56], [60], [61]) as well as future scientific investigations in the complex field of an ultra-wideband MIMO sensing and localizations.

7. Summary

Electromagnetic sounding for non-destructive and remote sensing, respectively, has been exploited for a long time. However, its practical application was mostly restricted to narrow-band sensors or it was banned to the laboratory in the case of wideband examinations. The reason for this limitation has been the lack of reasonable wideband measurement equipment.

The first field deployable ultra wideband devices were used in ground-penetrating radar (GPR). They mostly exploited powerful nanosecond or sub-nanosecond pulses to feed the transmission antenna. Meanwhile, several other UWB-sensor techniques have been introduced. Section 2 summarizes the most popular of them. The challenges of corresponding research and development are mainly to be seen in the performance improvement of the sensor electronics and its monolithic integration aimed at cost and power reduction.

The main part of the chapter deals with a pseudo-noise UWB approach and its main components. The pseudo-noise concept is an interesting alternative to other wideband sensing principles promoting both high device performance and monolithic integration. Due to its simple and rigid synchronization, it provides exact and time-stable signal generation and signal capture which promotes:

- simple adaptation of bandwidth, signal duration (period duration) and recording time to the needs of the actual application,
- the implementation of large MIMO-arrays,

- data processing in the time and frequency domain,
- device calibrations as usually with network analyzers,
- high-range precision and super-resolution capabilities, and
- excellent micro Doppler performance.

The most relevant RF components of a pseudo-noise sensor cover the test signal generation (i.e. pseudo noise code), the analog handling of the receive signals, and the high-speed conversion of the analog signals to the digital domain. Device concepts suited for these tasks are discussed in sections 3 to 5. Due to special requirements set by the application and the applied semiconductor technology, innovative solutions are presented. Among those are a distributed power amplifier with a novel cascode gain cell, new subtraction amplifiers, an analog-to-digital converter with a new reference network, and a high-speed predictor. Also, appropriate verification schemes are presented. A final section referring to implemented devices as they were applied in other UoKoLoS-projects suggests some first steps toward a fully integrated pseudo-noise sensor device.

Author details

Stefan Heinen, Ralf Wunderlich and Markus Robens
RWTH Aachen University, Germany

Jürgen Sachs and Martin Kmec
Ilmenau University of Technology, Germany

Robert Weigel, Thomas Ußmüller, Benjamin Sewiolo and Mohamed Hamouda
Friedrich-Alexander Universität Erlangen, Germany

Rolf Kraemer, Johann-Christoph Scheytt and Yevgen Borokhovych
Brandenburgische Technische Universität Cottbus, Germany

Acknowledgement

This work has been funded by the German Research Foundation (DFG) in the framework of the priority program UKoLoS (SPP 1202), project acronym HaLoS. The authors greatly appreciate the discussions and valuable support in electronic design given by M. Grimm, R. Herrmann, P. Rauschenbach and K. Schilling, as well as the support given by K. Borkowski in the manufacture of the electronic and mechanical components.

8. References

[1] J. Sachs, "Ultra-Wideband Sensing: The Road to new Radar Applications," in *IRS 2010 International Radar Symposium*, Vilnius (Lithuania), 2010.
[2] J. Sachs, *Handbook of Ultra-Wideband Short-Range Sensing - Theory, Sensors, Applications.* Berlin: Wiley-VCH, 2012.

[3] E. J. Candes, and M. B. Wakin, "An Introduction To Compressive Sampling," *Signal Processing Magazine, IEEE,* vol. 25, no. 2, pp. 21-30, 2008.

[4] E. J. Candes, "Compressive sampling," in *International Congress of Mathematicians,* Madrid, Spain, 2006.

[5] J. Ender, "Do we still need Nyquist and Kotelnikov? - Compressive Sensing applied to RADAR," in *8th European Conference on Synthetic Aperture Radar EUSAR 2010,* Aachen, Germany, 2010.

[6] H. Qiong, Q. Lele, W. Bingheng *et al.,* "UWB Through-Wall Imaging Based on Compressive Sensing," *Geoscience and Remote Sensing, IEEE Transactions on,* vol. 48, no. 3, pp. 1408-1415, 2010.

[7] P. v. Genderen, "Multi-Waveform SFCW radar," in *33rd European Microwave Conference,* 2003, pp. 849-852.

[8] S. Foster, "Impulse response measurement using Golay codes," in *IEEE International Conference on Acoustics, Speech, and Signal Processing,* ICASSP '86., pp. 929-932.

[9] J. Lee and J. D. Cressler, "Analysis and Design of an Ultra-Wideband Low-Noise Amplifier Using Resistive Feedback in SiGe HBT Technology," in *IEEE Trans. Microw. Theory Tech.,* Vol. 54, No. 3, pp. 1262-1268, March 2006.

[10] M.-C. Chiang et al., "Analysis, Design, and Optimization of InGaP-GaAs HBT Matched-Impedance Wide-Band Amplifiers with Multiple Feedback Loops," in *IEEE J. Solid-State Circuits,* Vol. 37, No. 6, pp. 694-701, June 2002.

[11] A. Ismail and A. A. Abidi, "A 3-10-GHz Low-Noise Amplifier With Wideband LC-Ladder Matching Network," in *IEEE J. Solid-State Circuits,* Vol. 39, No. 12, pp. 2269-2277, December 2004.

[12] F. Thiel, O. Kosch, and F. Seifert, "Ultra-Wideband Sensors for Improved Magnetic Resonance Imaging, Cardiovascular Monitoring and Tumour Diagnostics," in *Sensors 10,* pp. 10778-10802, December 2010.

[13] P. R. Gray et al., "Analysis and Design of Analog Integrated Circuits," 5th edit., John Wiley & Sons, 2010.

[14] M. Robens, R. Wunderlich, and S. Heinen, "UWB LNAs for Ground Penetrating Radar," in *IEEE Int. Symp. on Circuits and Systems,* Taipei, Taiwan, pp. 229-232, May 2009.

[15] H. Knapp et al., "15 GHz Wideband Amplifier with 2.8 dB Noise Figure in SiGe Bipolar Technology," in *IEEE Radio Freq. Integr. Circuits Symp.,* pp. 287-290, 2001.

[16] B. Kang et al., "Design and Analysis of a Cascode Bipolar Low-Noise Amplifier With Capacitive Shunt Feedback Under Power-Constraint," in *IEEE Trans. Microw. Theory Tech.,* Vol. 59, No. 6, pp. 1539-1551, June 2011.

[17] P. K. Datta and G. Fischer, "An Ultra-Wideband Low Power Consumption Differential Low Noise Amplifier in SiGe:C BiCMOS Technology," in *IEEE Radio and Wireless Symp.,* pp. 107-110, January 2006.

[18] M. Robens et al., "Differential UWB-LNA for M-Sequence Radar," in *Semicond. Conf. Dresden,* April 2009.

[19] M. Robens, R. Wunderlich, and S. Heinen, "Differential Noise Figure Measurement: A Matrix Based Approach," in *IEEE MTT-S Int. Microw. Symp. Dig.*, Anaheim, CA, pp. 385-388, May 2010.

[20] M. Robens, R. Wunderlich, and S. Heinen, "Differential Noise Figure De-Embedding: A Comparison of Available Approaches," in *IEEE Trans. Microw. Theory Tech.*, Vol. 59, No. 5, pp. 1397-1407, May 2011.

[21] H. Bosma, "On the Theory of Linear Noisy Systems," in Ph. D. dissertation, Dept. Elect. Eng., Eindhoven Univ. Technology, 1967.

[22] J. Randa, "Noise Characterization of Multiport Amplifiers," in *IEEE Trans. Microw. Theory Tech*, Vol. 49, No. 10, pp. 1757-1763, October 2001.

[23] A. A. Abidi and J. C. Leete, "De-Embedding the Noise Figure of Differential Amplifiers," in *IEEE J. Solid-State Circuits*, Vol. 34, No. 6, pp. 882-885, June 1999.

[24] J. Dunsmore and S. Wood, "Vector Corrected Noise Figure and Noise Parameter Measurements of Differential Amplifiers," in *European Microwave Conf.*, Rome, Italy, pp. 707-710, September 2009.

[25] S. W. Wedge and D. B. Rutledge, "Noise Waves and Passive Linear Multiports," in *IEEE Microw. and Guided Wave Letters*, Vol. 1, No. 5, pp. 117-119, May 1991.

[26] V. A. Monaco and P. Tiberio, "Computer-Aided Analysis of Microwave Circuits," in *IEEE Trans. Microw. Theory Tech.*, Vol. 22, No. 3, pp. 249-263, March 1974.

[27] M. Robens, R. Wunderlich, and S. Heinen, "Input Amplifier for Sensitivity Improvement in an M-Sequence Radar Front-End," in *Int. Conf. Indoor Positioning and Indoor Navigation*, pp. 450-451, September 2010.

[28] M. G. Chen and J. K. Notthoff, "A 3.3–V 21– Gb/s PRBS Generator in Al-GaAs/GaAs HBT Technology," *IEEE J. solid-State Circuits*, vol. 35, no. 09, pp. 1266–1270, Sept. 2000.

[29] T. Kjellberg, J. Hallin, and T. Swahn, "104 Gb/s 211-1 and 110 Gb/s 29-1 PRBS Generator in InP HBT Technology," in *IEEE International Solid-State Circuits Conference. Digest of Technical Papers*. 2004, Feb. 2006, pp. 2160–2169.

[30] H. Knapp, M. Wurzer, W. Perndl, K. Aufinger, J. Bock, and T. F. Meister, "100–Gb/s 27-1 and 54– Gb/s 211-1 PRBS Generators in SiGe Bipolar Technology," *IEEE J. Solid-State Circuits*, vol. 40, no. 10, pp. 2118–2125, Oct. 2005.

[31] T. O. Dickson, E. Laskin, I. Khalid, R. Beerkens, X. Jingqiong, B. Karajica, and S. Voinigescu, "An 80 Gb/s 231-1 Pseudorandom Binary Sequence Generator in SiGe BiCMOS Technology," *IEEE J. Solid-State Circuits*, vol. 40, no. 12, pp. 2735–2745, Dec. 2005.

[32] E. Laskin and S. P. Voinigescu, "A 60mW per Lane, 4×, 23– Gb/s 27-1 PRBS Generator," *IEEE J. Solid-State Circuits*, vol. 41, no. 10, pp. 2198–2208, Oct. 2006.

[33] F. Weiss, H.-D. Wohlmuth, D. Kehrer, and A. L. Scholtz, "A 24- Gb/s 27-1 Pseudo Random Bit Sequence Generator IC in 0.13 μm Bulk CMOS." Proceedings of the 32nd *European Solid-State Circuits Conference*, ESSCIRC 2006, p. 468-471

[34] H. Veenstra, E. van der Hejden, and D. van Goor, "A 15–27GHz Pseuo-Noise UWB Transmitter for Short-Range Automotive Radar in a Production SiGe Technology," in Proceeding of the *31st European Solid-State Circuits Conference*, 2005., Sept. 2005, pp. 275–278.

[35] J. P. Fraysse, J. P. Viaud, M. Campovecchio, P. Auxemery, and R. Qur, "A 2W, High Efficiency, 2-8GHz, Cascode HBT MMIC Power Distributed Amplifier," in *2000 IEEE MTT-S Int. Microwave Symp. Dig.*, vol. 1, June 2000, pp. 529–532.

[36] R. Zetik, J. Sachs, and R. S. Thomae, "UWB Short Range Radar Sensing," *Instrumentation and Measurement Magazine*, IEEE, vol. 10, pp. 39–45, Apr. 2007.

[37] J. Sachs, R. Herrmann, M. Kmec, M. Helbig, and K. Schilling, "Recent Advances and Applications of M-Sequence Based Ultra-Wideband Sensors." *IEEE International Conference on Ultra-Wideband*, ICUWB 2007.

[38] U. Schwarz, M. Helbig, J. Sachs, R. Stephan, and M. A. Hein, "Design and Application of Dielectrically Scaled Double-Ridged Horn Antennas for Biomedical UWB Radar Applications," in *IEEE International Conference on Ultra-Wideband*, ICUWB 2009., Sept. 2009, pp. 150–154.

[39] B. Sewiolo, J. Rascher, G. Fischer, and R. Weigel, "A 30 GHz bandwidth driver amplifier with high output voltage swing for ultra-wideband localization- and sensor systems," in *Proc. Asia-Pacific Conf.* (APMC'08), Dec. 2008, pp. 1–4.

[40] S. Trotta, H. Knapp, K. Aufinger, T. F.Meister, J. Bock, B. Dehlink, W. Simburger, and A. L. Scholtz, "An 84 GHz bandwidth and 20 dB gain broadband amplifier in SiGe bipolar technology," *IEEE J. Solid-State Circuits*, vol. 42, no. 10, pp. 2099–2106, Oct. 2007.

[41] C. Schick, T. Feger, E. Sonmez, K.-B. Schad, A. Trasser, and H. Schumacher, "Broadband SiGe HBT amplifier concepts for 40 Gbit/s fibreoptic communication systems," in *Proc. 35th Eur. Microw. Conf.*, Oct. 2005, vol. 1, pp. 113–116.

[42] Borokhovych Y. High-Speed Data Capturing Components for Super Resolution Maximum Length Binary Sequence UWB Radar. Brandenburgischen Technischen Universität; 2011. Available from: http://systems.ihp-microelectronics.com/web/-index.php5?id=16.

[43] Razavi B. Principles of Data Conversion System Design. IEEE Press; 1995.

[44] Analog-Digital-Umsetzer mit breitbandigem Eingangsnetzwerk. DE102009002062A1 07.10.2010; 2010.

[45] Kester W. Analog-Digital Conversion. ADI Central Applications Department; 2004.

[46] Borokhovych Y, Gustat H, Scheytt C. 4-bit, 16 GS/s ADC with new Parallel Reference Network. In: Proc. *IEEE Int. Conf. Microwaves, Communications, Antennas and Electronics Systems* COMCAS; 2009. p. 1–4.

[47] Borokhovych Y, H Gustat. 4-bit, 15 GS/s ADC in SiGe. In: *Proc. NORCHIP*; 2008. p. 268–271.

[48] Streng, B.: "Entwicklung und Implementierung eines Messplatzes zur Charakterisierung des HF-Teils von M-Sequenzmessköpfen in MATLAB," TU Ilmenau, Fak. EI, FG EMT, diploma thesis, 2009

[49] M. Kmec, M. Helbig, R. Herrmann, P. Rauschenbach, J. Sachs, K. Schilling, "M-Sequence Based Single Chip UWB-Radar Sensor," *ANTEM/AMEREM 2010 Conference*, July 5 – 9, 2010, Ottawa, ON, Canada.

[50] R. Herrmann, "Development of a 12 GHz bandwidth M-sequence based ultra-wideband radar and its application to crack detection in salt mines", TU Ilmenau, dissertation, 2011

[51] R. Herrmann, J. Sachs, K. Schilling, and F. Bonitz, "New extended M-sequence ultra wideband radar and its application to the disaggregation zone in salt rock," Proc. *12th International Conference on Ground Penetrating Radar*. Birmingham, UK, Jun. 2008.

[52] FCC News Release, "New Public Safety Applications and Broadband Internet Access Among Uses Envisioned by FCC Authorization of Ultra-Wideband Technology", 14 Feb 2002

[53] FCC 02-48, "Revision of Part 15 of the Commission's Rules Regarding Ultra-Wideband Transmission Systems", First Report & Order, Washington DC, Adopted 14 Feb 2002, Released 22 April 2002

[54] Electronic Communication Committee, "ECC Decision of 24 March 2006 amended 6 July 2007 at Constanta on the harmonized conditions for devices using Ultra ⊛Wideband (UWB) technology in bands below 10.6 GHz, " July, 6, 2007

[55] M. Kmec, J. Sachs, P. Peyerl, P. Rauschenbach, R. Thomä, R. Zetik, "A Novel Ultra-Wideband Real-Time MIMO Channel Sounder Architecture," *XXVIIIth URSI General Assembly 2005*, Oct. 23-29, New Delhi, October 2005.

[56] M. Helbig, I. Hilger, M. Kmec, G. Rimkus, J. Sachs, "Experimental phantom trials for UWB breast cancer detection," *German Microwave Conference*, GeMiC 2012, Ilmenau

[57] M. Lazebnik, E.L. Madsen, G.R. Frank et al.: „Tissue-mimicking phantom materials for narrowband and ultrawideband microwave applications" *Physics in medicine and biology* , vol. 50, 2005, 4245-4258

[58] E. C. Fear, S. C. Hagness, P. M. Meaney, M. Okoniewski, and M. A. Stuchly, "Enhancing breast tumor detection with near-field imaging," *IEEE Microwave Magazine*, vol. 3, no. 1, pp. 48–56, Mar. 2002

[59] I.J. Klemm J.A. Craddock, A. Leendertz et al.: "Radar-based breast cancer detection using a hemispherical antenna array – experimental results," *IEEE Trans on Antennas and Propagation*, vol. 57, 2009, 1692-1704

[60] Kosch O., Thiel F., Ittermann B., and Seifert F., "Non-contact cardiac gating with ultra-wideband radar sensors for high field MRI", *Proc. Intl. Soc. Mag. Reson. Med. 19*, (ISMRM), Montreal, Canada, ISSN 1545-4428, p.1804, (2011)

[61] M. Helbig, M. Kmec, J. Sachs, C. Geyer, I. Hilger, G. Rimkus, "Aspects of antenna array configuration for UWB breast imaging Brust", *6th European Conference on Antennas and Propagation*, Prague, March 2012

[62] Afzali-Kusha, A.; Nagata, M.; Verghese, N.K.; Allstot, D.J.; , "Substrate Noise Coupling in SoC Design: Modeling, Avoidance, and Validation," *Proceedings of the IEEE* , vol.94, no.12, pp.2109-2138, Dec. 2006

[63] S.M. Sze, *Physics of Semiconductor Devices 2nd edn*, New York, Wiley, 1981

[64] J.A. Olmstead and S. Vulih, "Noise problems in mixed analog-digital integrated circuits," in Proc. *IEEE Custom Integrated Circuits Conf.*, 1987, pp. 659-662

UWB in Medicine – High Performance UWB Systems for Biomedical Diagnostics and Short Range Communications

Dayang Lin, Michael Mirbach, Thanawat Thiasiriphet, Jürgen Lindner, Wolfgang Menzel, Hermann Schumacher, Mario Leib and Bernd Schleicher

Additional information is available at the end of the chapter

1. Introduction

This chapter presents scientific achievements in the field of UWB radar and communication systems for biomedical applications. These contributions focus on low-power MMIC designs, novel antenna structures and competitive approaches for communication and imaging.

The first section describes components for UWB radar sensors and communication systems, namely antennas and integrated circuits. Novel broadband antenna concepts for UWB radar and communication applications are presented. Symmetrical UWB antenna structures for free space propagation with improved performance compared to existing antennas regarding radiation pattern stability over frequency are designed, realized and successfully characterized. Novel differential feeding concepts are applied, suppressing parasitic radiation by cable currents on feed lines. For applications such as communication with implants and catheter localization, a miniaturized antenna optimized for radiation in human tissue is designed. The radiation characteristics of the antenna are measured using an automated setup embedded in a liquid consisting of sugar and water, mimicking the dielectric properties of biological tissue. For UWB radar transmitters, a differential and low-power impulse generator IC is realized addressing the FCC spectral mask based on a quenched cross-coupled LC oscillator. The total power consumption is only 6 mW at an impulse repetition rate of 100 MHz. By adding a simple phase control circuit setting the start-up phase condition of the LC oscillator, an impulse generator with a bi-phase modulation scheme is achieved. A further modification introduces a variable width of the pulse envelope as well as a variable oscillation frequency. The corresponding spectra have controllable 10 dB bandwidths and center frequencies fitting the different spectral allocations in the USA, Europe and Japan. On the receiver side, both a fully differential correlation-based and an energy detection receiver for the 3.1-10.6 GHz band are designed. Monostatic UWB radar systems require transmit/receive turn-around times in the nanosecond regime. Integrated front-ends which successfully address this issue are presented here for the first time.

The second section deals with signal processing. As, due to the large RF bandwidth, direct analog to digital conversion and digital signal processing are not feasible (at least not at reasonable power consumption), analog signal processing is one focus. For communication, detection methods based on analog correlation require channel estimation, storing of impulse responses and also precise time synchronization. Therefore methods based on energy detection are developed which require no or little channel knowledge, having low complexity, robustness to multipath propagation and high resistance to synchronization and symbol clock errors. New modulation techniques are described, which can cope with interchip and intersymbol interference. Also a novel support by a comb filter resulting in significant SNR improvements in interference and multiuser scenarios is presented. The methods developed for communication applications can also be used in the radar context. For detection and tracking of moving targets (e.g. heart in the body) new algorithms based on particle filtering are developed for the digital signal processing part. It is shown that the accuracy, the resolution and robustness can be improved compared to conventional methods. For the objective of catheter localization, the knowledge of the shape and position of the human body surface is inevitable. A UWB imaging algorithm for the detection and estimation of this surface has been developed based on trilateration and is also described in this second section. Furthermore, building on this surface estimation algorithm, a new method for the localization of transmitters in dielectric media is presented. Taking into account the refraction effects on the boundary surface, the algorithm uses the impulse time of arrival to determine the transmitter position inside of the dielectric medium.

The third section finally describes the design of bistatic UWB radar systems using the components presented in the first section. Single-ended and differential radar demonstrators are developed, with which the potential of impulse-radio UWB sensing is evaluated. Measurements aimed at applications of the developed hardware such as vital sign monitoring and communication with implants are presented. Further measurements are performed to prove the functionality of the imaging algorithms derived in the second section. For surface estimation, a single radar sensor is moved around a highly reflective target in order to emulate a whole sensor array. For the verification of subsurface transmitter localization, a transmitter is placed inside of a container filled with tissue mimicking liquid, and its position is visualized with respect to the estimated container surface.

2. Circuit and component design

2.1. UWB antennas

Concepts for antennas with an ultra-wideband behavior are well-known and established [30]. However, advancements focusing on specific applications and specific performance parameters are inevitable to keep pace with the requirements of modern communication, radar, and localization systems. In particular in the medical environment, application-oriented antennas are mandatory to cope with the need for reliable systems (e. g. health monitoring systems) or with extreme environmental conditions (e. g. implanted systems). In the following, three novel UWB antennas are introduced targeting different tasks in the medical field and showing outstanding characteristics with respect to certain key antenna parameters.

2.1.1. Circular slot antenna excited with a dipole element

Figure 1. Dipole slot antenna.

Planar broad monpoles or dipoles are favored UWB antennas for communication systems with high data rates, e. g. potentially used in base stations for patient monitoring. However, broad monopoles fed single-endedly are prone to cable currents on the feeding line disturbing the radiation characteristic in the lower frequency range [15], while dipoles behave like a λ-radiator with a zero in main beam direction in the upper frequency range. Both effects lead to an undesired change of the radiation pattern in the operational frequency range reducing the effective bandwidth. For the widely-used impulse based UWB systems, this leads to a broadening of the impulse and consequently to a degraded system performance.

A practical solution to overcome the described parasitic radiation pattern performance is the combination of a circular slot antenna with a dipole feeding element as depicted in Fig. 1. The circular slot behaves like a broad monopole according to Babinet's principle. A broad dipole located in the center of the circular slot and consisting of two circular segments excites the slot antenna. The inherent symmetrical feeding of the dipole avoids the propagation of cable currents due to the virtual ground plane in between the transmission lines and results in an uniform radiation characteristic over the UWB frequency range.

The length of the exciting dipole is designed to be $\lambda/2$ at the center of the FCC UWB frequency range. Therefore, the dipole is smaller than pure UWB dipole antennas with a typical length of $\lambda/2$ at the lower edge of the FCC UWB frequency range. The perimeter of the circular slot is about λ at the lower edge of the FCC UWB frequency range leading to a resonance at 4.3 GHz (see simulation result for $|S_{11}|$ in Fig. 2(a)), and hence, to a return loss better than 10 dB at 3.1 GHz. Additional resonances with a low qualtiy factor are arising if the perimeter of the slot is a multiple of the wavelength (see at 6.9 GHz and 9.8 GHz in Fig. 2(a)). Therefore, a UWB behavior regarding return loss is achieved.

In order to characterize the dipole slot antenna with a single-ended coaxial line, a common UWB planar transition from coplanar stripline to a microstrip line based on [32] is used. A metallic shielding around this balun suppresses any parasitic radiation (see Fig. 1). The

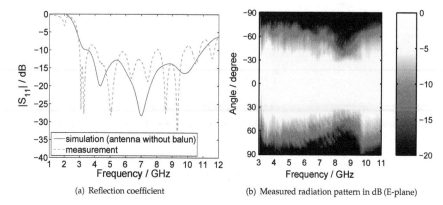

(a) Reflection coefficient   (b) Measured radiation pattern in dB (E-plane)

Figure 2. Reflection and radiation performance of the dipole slot antenna.

measured return loss behavior of the antenna including balun is presented in Fig. 2(a) and shows the predicted UWB performance.

The measured radiation pattern demonstrates the desired uniform characteristic over the FCC frequency range, as can be seen in Fig. 2(b) for the E-plane. In the H-plane a similar pattern is obtained having a slightly narrower beam with a mean 3 dB beamwidth of 60° in contrast to 65° for the E-plane. The uniform radiation behavior in the frequency domain results in a short antenna impulse response making the antenna a suitable candidate for impulse-based communication systems.

An upgrade of this single polarized antenna for horizontal and vertical polarizations is possible, if two narrow feeding dipoles are placed in orthogonal position to each other inside the circular slot. The expense of the additional feature is a deterioration in the return loss and radiation characteristic (see [20]).

2.1.2. Dielectric rod antenna fed by a planar circular slot

Emerging applications in medicine are vital sign monitoring [33], breast cancer detection [11], and tracking of inner organs for improved magnetic resonance tomography [38]. For these radar based sensing systems, directive antennas with a small beamwidth in both planes are compulsory. A promising approach to achieve this attribute is to exploit dielectric rod antennas either fed by a tapered slot antenna [14] or by a biconical dipole [4]. Both ideas show very good electrical characteristics, but suffer from necessary sophisticated fabrication and assembly.

A new antenna proposal is shown in Fig. 3(a). In this concept, the planar circular slot antenna presented in Sec. 2.1.1 acts as feed for a circular dielectric waveguide. The electrical field distribution inside the circular slot is similar to the fundamental mode H_{11} of the dielectric waveguide. Hence, this mode is predominantly excited. Due to that and since the H_{11}-mode possesses no cutoff frequency, in general an ultra-wideband performance is obtained. The diameter of the dielectric rod is chosen to 43 mm, which is a compromise between a good return loss behavior and single-mode operation of the dielectric waveguide. The conical shape at the end of the rod acts as smooth transition of the waveguide impedance to the free space impedance.

Due to the rod permittivity of 2.8, the electromagnetic field is mainly concentrated within the rod. Therefore, the antenna primarily radiates in forward direction along the rod. The unidirectionality and, hence, the gain is slightly improved – especially for lower frequencies – placing a metallic reflector at the backside of the antenna (see Fig. 3(a)). The reflector distance for the shown structure is $\lambda/4$ at 5.35 GHz leading to an optimized mean gain.

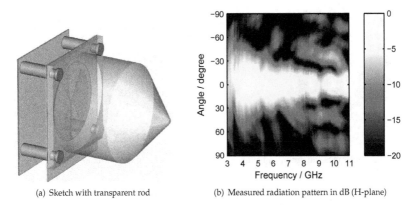

(a) Sketch with transparent rod   (b) Measured radiation pattern in dB (H-plane)

Figure 3. Sketch and radiation performance of the dielectric rod antenna fed by the dipole slot antenna.

The radiation pattern in Fig. 3(b) shows the obtained small beamwidth of the antenna for the H-plane. The increased side lobe level at frequencies above 8 GHz is caused by a parasitic leaky wave and is hardly avoidable for rod antennas. The given radiation characteristics in Fig. 3(b) are typical for both planes and results in a high mean gain of 8.7 dBi including the return loss. Due to the fact that the return loss is better than 10 dB from 3.5 GHz to 11.8 GHz, the influence of the return loss on the realized gain is negligible.

Besides this good electrical performance, the major benefits of this antenna in contrast to [14] and [4] are the compactness, the low weight, and the ease of fabrication. These attributes make the antenna also interesting for all kinds of industrial applications.

2.1.3. UWB slot antenna optimized for radiation in human tissue

High data rate communication for implanted devices [5] or precise catheter localization inside the human body are futuristic topics in medicine. There, impulse-based UWB technology is advantageous compared to narrow band systems due to the low power consumption caused by the simple system architecture. However, UWB antennas optimized for radiation in human tissue are hardly investigated.

In Fig. 4(a) a UWB antenna is proposed for radiation in human tissue, which is based on a similar concept as the dipole slot antenna in Sec. 2.1.1. An elliptical slot antenna is fed by a broad monopole located in the center of the slot. A monopole is chosen to obtain a small structure and to avoid a bulky balun for characterization purposes. Instead of a single layer structure, two substrate layers with slots in the top and bottom layer metalizations are used. The monopole is arranged in the center metalization layer and is fed by a triplate line. In this way, the buried feeding is insulated from the adjacent highly lossy human tissue. The antenna dimensions need to be optimized according to the surrounding medium of the antenna. The width of the antenna is 11 mm assuming skin tissue around the antenna with a typical permittivity of 28 at 6.85 GHz. The size reduction factor compared to an antenna designed for air instead of skin tissue is 5.4 leading automatically to a miniaturized UWB antenna.

(a) Sketch

(b) Photograph

Figure 4. Sketch of the tissue optimized antenna and photograph including microstrip-to-coaxial transition.

In order to connect the antenna to a coaxial cable, a broadband transition from the triplate line to a coaxial line is applied [19]. The realized antenna including this transition is depicted in Fig. 4(b). The characterization of the tissue optimized antenna is performed in a tank filled with tissue-mimicking liquid approximating the permittivity and loss behavior of skin tissue. The chosen liquid is a 50% sugar solution in water [20]. Fig. 5(a) shows the return loss of the antenna being inside the sugar-water solution and compares its performance with a measurement, where the antenna is placed on both sides on human skin. Both measurement results agree very well and show a return loss of more than 10 dB above 3.8 GHz.

The radiation pattern is also measured in the tissue-mimicking liquid using two identical antennas and applying the two-antenna method. The obtained radiation pattern for the H-plane is presented in Fig. 5(b). There, the losses of the tissue-mimicking medium are compensated. Since the losses are increasing significantly with increasing frequency, measurements only up to 9 GHz are possible limited by the dynamic range of the measurement setup. Within this frequency range, a desired uniform and broad characteristic is achieved. Hence, UWB performance for a sufficiently small antenna for implants is demonstrated. For catheter localization additional miniaturization is required. A possible approach as well as more details about all presented antennas in Sec. 2.1 can be found in [20].

2.2. Transmitter MMICs

All integrated circuits reported in this section were realized in an inexpensive Si/SiGe HBT technology offered by Telefunken Semiconductors GmbH. Two kinds of transistors, with high f_T ($f_T = 80$ GHz, $BV_{CEO} = 2.4$ V) and with high breakdown voltage ($f_T = 50$ GHz, $BV_{CEO} = 4.3$ V) are available simultaneously. The process incorporates 4 types of resistors, MIM capacitors, as well as 3 metalization layers. All the devices were fabricated on a low resistivity 20 Ω cm

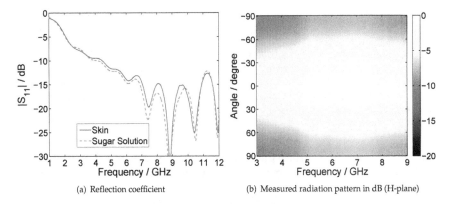

(a) Reflection coefficient

(b) Measured radiation pattern in dB (H-plane)

Figure 5. Reflection and radiation performance of the tissue optimized antenna.

substrate. The technology is fully adequate for impulse-radio-ultra-wideband (IR-UWB) applications.

Generating short time-domain impulses making efficient use of the spectral mask is the key challenge in IR-UWB systems. Approaches include the up-conversion of base band pulses to the allocated UWB frequency band using an oscillator and mixer [39] and direct generation based on damped relaxation oscillator [8]. Here impulse generators based on a quenched-oscillator concept with great circuit simplicity are presented. A cross-coupled LC oscillator is chosen as the core to introduce tunability of the waveform and the inherent convenience of achieving biphase modulation.

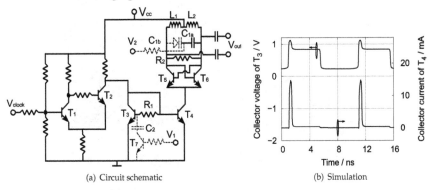

(a) Circuit schematic

(b) Simulation

Figure 6. Complete circuit schematic of the UWB impulse generator. The dashed components (C_{1b}, C_2 and T_7) show the extension for tunability of the impulse shape for different spectral masks and transistor level simulation: the collector potential of T_3 and the collector current of T_4.

Fig. 6(a) shows the impulse generator circuit. First, disregard the components with a dashed line, which are the extention for tunability of the impulse shape. T_1 and T_2 form a Schmitt trigger, creating a fast rising edge at the collector of T_2 when a positive input signal triggers T_1 to be on. This reduces the effect of the time-domain influence of the input clock signal on the output impulses. After an initial sharp increase, the collector potential of T_2 is pulled down again to a lower constant value by the current mirror formed by T_3 and T_4. So the collector potential of T_3 has a spike performance before it becomes stable, as shown in Fig. 6(b), which correspondingly generates a collector current spike at T_4, creating the envelope of the output impulse. During the rest of the period, the collector current of T_4 is too low to turn the oscillator on since T_3 is chosen much larger than T_4. The width of the current spike is determined predominately by the time constant τ of the charging circuit including the resistor R_1 and the base-emitter capacitor C_{be3} of T_3. τ can be written as

$$\tau = R_1 C_{be3}. \tag{1}$$

The amplitude of the current spike can be easily adjusted by changing the size of T_4. The repetition rate of the current spike train is equal to the input signal frequency and limited by the spike width.

The LC oscillator is activated by the current spikes once the collector currents of the cross-coupled pair (T_5 and T_6) are high enough to create a negative real part of the impedance. A slight asymmetry in the cross-coupled pair ensures that the oscillation always starts with the same phase and shortens the start-up time which in turn reduces power consumption because the necessary current spike width for a given output impulse envelope is shortened. R_2 is placed to quench the oscillator off more quickly immediately after the current spikes have disappeared. Thus, short-time domain impulses with a repetition rate equal to the input signal frequency are generated. The center frequency ω_0 of the oscillation is mainly determined by L_1, L_2, C_{1a} and the parasitic capacitance from the cross-coupled pair C_{para}. ω_0 can be expressed as

$$\omega_0 = \frac{1}{\sqrt{(C_{1a} + C_{para})(L_1 + L_2)}}. \tag{2}$$

It is designed to be around 6 GHz to fit the FCC spectral mask. The microphotograph of this realized impulse generator is shown in Fig. 7(a). It is a quite compact design with an area of $0.50 \times 0.60\,\text{mm}^2$ due to a simple circuit topology.

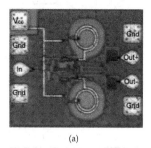 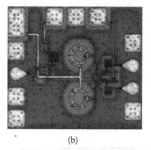

(a)   (b)

Figure 7. Microphotographs of (a) the realized UWB impulse generator shown in Fig. 6(a) excluding the components with a dashed outline and (b) the impulse generator tunable to FCC, ECC and Japanese spectral masks.

The impulses measured on-chip in time domain are shown in Fig. 8(a). The impulse generator is fed with a 100 MHz and 1.3 GHz sinusoidal signal separately. The differential signal is

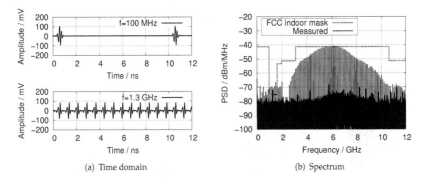

(a) Time domain   (b) Spectrum

Figure 8. Measured results of time-domain output impulse waveforms at 100 MHz and 1.3 GHz repetition rate and the spectrum of the 100 MHz impulse train, demonstrating compliance with the FCC indoor spectral mask. The impulse generator is shown in Fig. 7(a)

displayed using the subtraction feature of a real time oscilloscope. The measured results show a peak-peak amplitude of 200 mV, and a full width at half maximum (FWHM) of the envelope of 0.3 ns. This circuit has a very low power consumption: 6 mW at 100 MHz and 10 mW at 1.3 GHz. The spectrum of the measured 100 MHz impulse train can be seen in Fig. 8(b). The maximum power spectral density (PSD) is -41.3 dBm per spectral line, and it has a 10 dB bandwith of 4.9 GHz from 3.5-8.4 GHz. It shows that the output spectrum complies well with the FCC mask for indoor UWB applications.

The two parameters (τ and ω_0) which determine the impulse parameters (envelope and osillation frequency) are easily modified. This is shown by the dashed components in Fig. 6(a). C_2 and T_7 are introduced to modify the envelope by changing the capacitance between the base of T_3 and ground, switching C_2 in and out. When $V_1 = 0$, T_7 is off, τ is the same as before, resulting the emitted impulses to conform to the FCC mask. When $V_1 = 1$ V, T_7 is on, the charging circuit will include R_1 and C_{be3} in parallel with C_2. In this case, the time constant τ_1 can be written as

$$\tau_1 = R_1(C_{be3} + C_2). \tag{3}$$

Since C_2 is chosen much larger than C_{be3}, the envelope width of the impulses is larger in this situation, suiting for ECC or Japanese masks depending on the center frequency adjustment. The tank circuit capacitance is now formed by C_{1a} in series with a varactor C_{1b}. The oscillation frequency can be expressed as

$$\omega_0 = \frac{1}{\sqrt{(\frac{C_{1a}C_{1b}}{C_{1a}+C_{1b}} + C_{para})(L_1 + L_2)}}. \tag{4}$$

Through changing the varactor capacitance C_{1b} with V_2, the center frequency ω_0 is adjustable. Depending on the applied voltages V_1 and V_2, the generated impulses conform to the FCC, ECC mask, or Japanese mask. The microphotogragh of this tunable impulse generator can be seen in Fig. 7(b). It is quite compact with an area of 0.53 x 0.61 mm^2.

By setting $V_1 = 0$ and $V_2 = 2$ V, the LC oscillator is triggered with a shorter current spike. So the generated waveform is similar as shown in Fig. 8(a), targetting the FCC mask. This impulse generator is suitable for the ECC mask when $V_1 = 1\,V$ and $V_2 = 2.3\,V$. Under these conditions, the LC oscillator is triggered by a longer current spike with a FWHM of 2 ns. The measured output impulse train can be seen in Fig. 9(a). The impulses have a peak-peak amplitude of

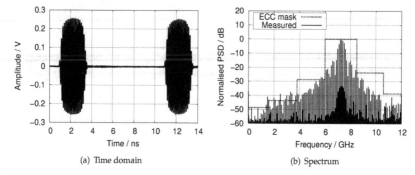

(a) Time domain   (b) Spectrum

Figure 9. Measured results of impulse waveform in time domain and normalized PSD of the tunable impulse generator output signal setting for compliance with the ECC UWB mask.

0.5 V. The circuit has a total power consumption of 10 mW and a maximum output impulse repetition rate exceeding 300 MHz in this case. The normalized PSD of the impulse train has a center frequency around 7 GHz with a 10 dB bandwidth of 600 MHz, shown in Fig. 9(b). It fits well into the ECC mask. By changing the value of V_2, the center frequency of the impulses will be shifted, this makes the circuit usable for the Japanese mask. The measured output signal

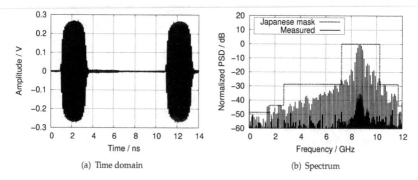

(a) Time domain   (b) Spectrum

Figure 10. Measured results of time domain waveform and normalized PSD of the output signal targeting the Japanese UWB mask.

in the time domain with $V_{tune2} = 6\,V$ is shown in Fig. 10(a). The measured impulse train has a similar envelope as the mode targeting the ECC mask because the triggered current spike has the same width. The peak-peak amplitude of the impulses whose envelope has a FWHM of 2 ns is 0.5 V. The complete power consumption in this mode is 10 mW. The spectrum is presented in Fig. 10(b). It shows that the center frequency is shifted to 8.7 GHz for a good fit to the Japanese mask.

The performance under these three modes is summarized in Tab. 1. This impulse generator can be used for on-off keying (OOK) and pulse-position modulation (PPM) in all these three modes.

Mode (setting)	10dB bandwidth (GHz)	V_{PP} (V)	power cons. (mW)
FCC $V_1=0$, $V_2=2.0$ V	4.2	0.36	6
ECC $V_1=1$ V, $V_2=2.3$ V	0.6	0.5	10
Japanese mask $V_1=1$ V, $V_2=6$ V	0.6	0.5	10

Table 1. Performance summary of the tunable impulse generator.

Biphase modulation capability can be introduced by modifying the DC currents flowing in the individual branches of the differential LC oscillator. As shown in Fig. 11(a), additional branch currents are set through current mirrors T_7, T_9 and T_8, T_{10}. When the input data signal is

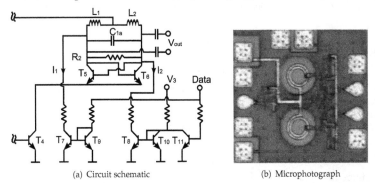

(a) Circuit schematic   (b) Microphotograph

Figure 11. Adding biphase capability to the impulse generator in Fig. 6 and the microphotograph of the impulse generator with biphase modulation function.

low, transistor T_9 is off, and the collector current I_1 of T_7 is zero because of the current mirror configuration to T_9. Meanwhile, the applied voltage V_3 will generate a collector current I_2 in T_8 through the current mirror configuration of T_8 and T_{10}, because T_{11} is off. When the data signal is high, T_{10} switches into saturation and T_8 blocks, which causes I_2 to be zero. At the same time, the high potential at the base of T_9 introduces a collector current I_1 in T_7. Thus, oscillation will start in one of these two phase states once the tail transistor T_4 is turned on, constituting the biphase modulation. Additionally, this asymmetry shortens the start-up time, which in turn reduces the power consumption. The fabricated IC is shown in Fig. 11(b). It has an area of 0.56 X 0.53 mm².

The measured time domain waveforms with different voltage potentials applied to the data port can be seen in Fig. 12(a). The results show a peak-peak amplitude of 260 mV and an envelope width of 0.3 ns FWHM. The orientation of the impulses is clearly reversed, showing a perfect biphase modulation. Fig. 12(b) shows the spectrum of a 200 MHz impulse train with the data port connected to ground. It is centered around 6 GHz with a 10 dB bandwidth of

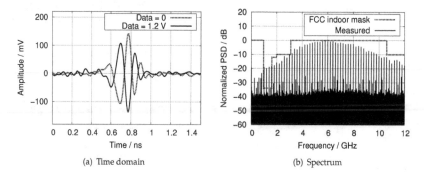

(a) Time domain   (b) Spectrum

Figure 12. Measured time-domain results of biphase modulated impulses with different applied voltages at the data port and the spectrum information of a 200 MHz impulse train with the data port connected to ground.

6.7 GHz from 3.1 - 9.8 GHz, which complies well with the FCC spectral mask for indoor UWB systems.

2.3. Receiver MMICs

Two different types of receivers will be described in this section. Energy detection receivers are implemented for on-off keying (OOK) communications and localization applications, they are conceptually simple and do not require syncronization, but are also sensitive to interference. Correlation detection receivers are introduced to solve this problem, they are more robust to interference, but require accurate timing synchronization with the transmitter. This problem is eliminated in radar applications, because transmitter and receiver are co-located and synchronized with a common reference.

A fully differential UWB low-noise amplifier is a key element for both receivers. The LNA should provide a low noise figure, a high gain, a flat frequency response, and a small group delay variation within the complete frequency range. Another key component is a four quadrant analog multiplier, which performs the squaring operation in the energy detection receiver and the multiplication operation in the correlation receiver. Detailed explanation of these components will be described below.

2.3.1. Fully differential UWB low-noise amplifier

Fig. 13(a) shows the fully differential UWB low-noise amplifier schematic. It consists of a differential cascode, followed by two emitter follower stages as buffers. Input and output are differential as the LNA will be connected to a symmetrical antenna, and shall feed a Gilbert cell type analog multiplier directly, without an unbal circuit. The symmetry of the emitter-coupled pair is achieved by placing identical transistors and passive components in the two branches.

T_1 through T_4 form the differential cascode which is biased by the stacked current mirror. The primary reason of the cascode configuration is to reduce the Miller effect at the input port, increasing the bandwidth. The shunt-shunt feedback (R_1, C_1 and R_2, C_2) further broadens

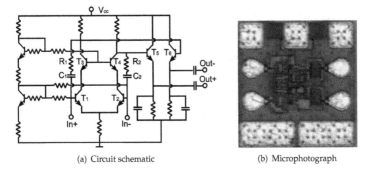

(a) Circuit schematic   (b) Microphotograph

Figure 13. Complete circuit schematic and the chip microphotograph of the fully differential low-noise amplifier.

the bandwidth and improves the input matching simultaneously. Careful selection of input transistor size and adjusting the bias point was done as a compromise between optimum current density for minimum noise figure, noise-matched input impedance and achievable bandwidth. The emitter size of T_1 is chosen to be 0.5 μm x 24.7 μm and the emitter current is 5 mA. The wide band noise and input power match were accomplished by the selection of input transistor with suitable biasing and shunt-shunt resistive feedback. A negligible penalty, with a maximum value of 0.2 dB, is achieved within the entire band for not achieving noise match exactly. T_5, T_6 form a differential emitter follower buffer. The emitter degeneration capacitors are used to improve the buffer bandwidth.

The microphotograph of this differential LNA is shown in Fig. 13(b). Because this design is completely inductor-less, the IC has an extremely small size of 0.37 X 0.38 mm^2 including all bound pads. The lowest available metal layer was placed below the large-sized bonding pads to provide a ground shield, as otherwise the noise figure may be deteriorated by the substrate noise pick-up.

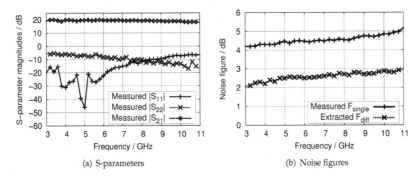

(a) S-parameters   (b) Noise figures

Figure 14. Measurement results of the S-parameter magnitudes and single-ended and extracted differential noise figures.

One drawback of the differential configuration is a complex measurement setup. Two identical passive microstrip line UWB baluns are used for differential S-parameter measurement. The influence of the baluns is removed during the calibration process. The measurement is restricted to 3 - 11 GHz due to the operating range of the UWB baluns. Fig. 14(a) shows the measured S-parameters. The non-ideal performance of the UWB baluns introduces ripples in the measured curves. The measurement results show a differential gain of 19.9 dB with a 1.8 dB variation, the input matching has a value of smaller than -7 dB and the output one is smaller than -6 dB in the complete FCC allocated frequency range. The method in [2] is adopted to extract the differential noise figure. First a single-ended noise figure F_{single} is measured from port In+ to Out- with the other ports terminated by 50 Ω resistors. Then, by measuring the transducer gain from port In+ to Out- (G_{31}) and Out+ to In+ (G_{32}), the differential noise figure can be extracted as

$$F_{diff} = 1 + \frac{1}{G_{31} + G_{32}} (F_{single} G_{31} - G_{31} - G_{32}). \tag{5}$$

Fig. 14(b) shows the information of the noise figures. The differential noise figure varies from 2 dB at 3 GHz to 2.9 dB at 10.6 GHz. Small group delay variation within the entire band is required for single-band IR-UWB systems. As depicted in Fig. 15(a), the group delay variation is smaller than 15 ps within the complete band. Fig. 15(b) shows the measured large signal behavior at 7 GHz of this differential amplifier. The input 1 dB compression point is -17.5 dBm. The complete power consumption of this differential LNA is 77 mW.

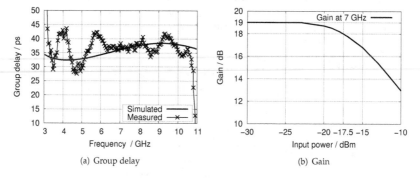

(a) Group delay

(b) Gain

Figure 15. Measured results of group delay versus frequency and gain depending on the input power.

2.3.2. Energy detection receiver

The core of energy detection receivers is a squaring circuit. Fig. 16(a) displays the squaring circuit based on a Gilbert cell four quadrant multiplier comprising two differential stages in parallel with cross-coupled output, complemented by a low-pass filter and a differential output buffer. The squaring operation is realized by connecting the same signal to both inputs of the Gilbert cell. The signal fed to the lower pair of the Gilbert cell is taken directly from the LNA output transistors, while the signal fed to the top quad is passed first through the emitter follower buffer. Both paths introduce almost the same group delay. Thus, the two branches of the input signal arrive simultaneously at the multiplier, ensuring an exact squaring operation. The load resistors (R_1, R_2) of the Gilbert cell, together with shunt capacitors (C_1, C_2) of the

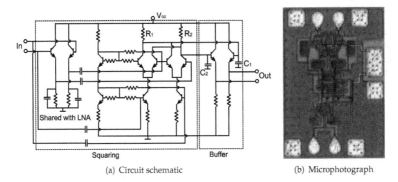

(a) Circuit schematic

(b) Microphotograph

Figure 16. Squaring circuit, low-pass filter and buffer of the energy detection receiver.

output buffer form low-pass filters with 1 GHz 3 dB bandwidth, which are needed for the envelope detection. The LNA from Fig. 13(a) is added to complete the energy detection receiver, which totally consumes 108 mW. Fig. 16(b) shows the microphotograph of the fabricated receiver IC, it measures 0.43 mm x 0.61 mm, including bond pads.

For testing the energy detection receiver, a 700 Mbit/s return-to-zero (RZ) impulse train was generated by the impulse generator shown in Fig. 6(a), which has a power comsumption of 7.5 mW at this rate. The transmitter and receiver ICs are seperately mounted on Rogers RO4003C substrates which also carry the dipole-fed circular slot antennas discussed in 2.1.1, and are wire-bonded to microstrip transmission lines feeding the antennas. The two antennas are placed at a distance of 30 cm. Fig. 17(a) shows the input data sequence from a pattern

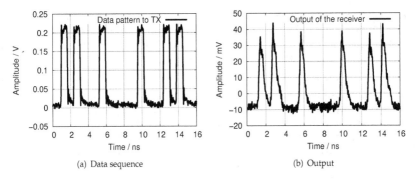

(a) Data sequence

(b) Output

Figure 17. OOK transmission experiment at 700 Mbit/s over 30 cm, data sequence (700 Mbit/s data rate) applied to the transmitter and detected signal at the receiver output.

generator. The corresponding detected impulse envelopes with a peak amplitude of 40 mV at the output of the receiver IC can be seen in Fig. 17(b). This experiment clearly demonstrates that the simple transmitter/receiver combination can be used to transmit significant bit rates over short distances. Detailed measured results of the receiver are shown in [23].

2.3.3. Correlation detection receiver

Coherent detection receivers are based on the cross-correlation realized by feeding the received signal and the on-chip generated template impulse into a wideband analog four-quadrant multiplier and subsequent low-pass filtering. Fig. 18(a) shows the block diagram of the correlation receiver. The multiplier-based correlation is done in the RF domain, which leads to an energy efficient solution by omitting power-hungry wideband ADCs. In a radar setup, the transmit and receive clocks need to be phase adjusted, which in practice is done by a DDS board. The complete schematic of the UWB analog correlator circuit

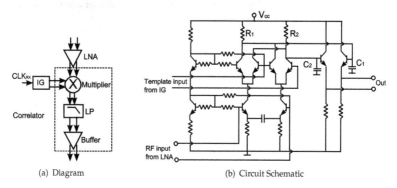

(a) Diagram   (b) Circuit Schematic

Figure 18. Architecture of the correlation receiver system and the schematic of the correlator with a true multiplier, a low-pass filter and a buffer.

can be seen in Fig. 18(b). The core of the correlator is again a Gilbert cell which acts as a wide-band multiplier with proper template impulse amplitude applied to the switching quad. Capacitively shunted resistive emitter degeneration results in the necessary gain flatness over the whole UWB frequency band. The low-pass filters are formed by the load resistors (R_1, R_2) of the Gilbert-cell with the shunt capacitors (C_1, C_2) of the buffer, the same as shown for the case of the energy detection above. The correlator consumes 35 mW.

The complete receiver IC, including the differential LNA, the correlator and the template impulse generator, is shown in Fig. 19(a). It measures 0.43×0.61 mm^2 and consumes a total DC power of 130 mW. To demonstrate the correlation performance of the receiver, the receiver IC was connected to a dipole-fed slot antenna, and placed at a distance of 20 cm from the transmitter. A small offset frequency of 100 Hz was introduced between the transmitter and receiver clocks, making the template impulses continuously sweep through the received signal. The measured cross-correlation can be seen in Fig. 19(b). More details of the correlation receiver can be found in [24].

2.4. Monostatic radar MMICs

All UWB radar sets reported so far use a bistatic antenna configuration. A monostatic UWB radar would significantly reduce the size of IR-UWB sensors because of the elimination of one antenna. However, implementation of rapid switching between the transmit and the receive path is difficult to realize in either this low-cost bipolar-only or CMOS technology. Here, a

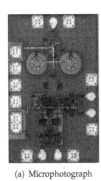

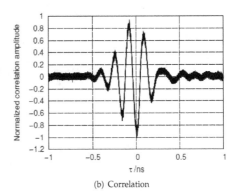

(a) Microphotograph (b) Correlation

Figure 19. Chip micrograph of the fully integrated correlation receiver and the measured normalized cross-correlation of the received impulse with the template impulse.

novel front-end concept based on a merged impulse generator/low noise amplifier, shown in Fig. 20(a) is proposed. In this design, the input of the differential low-noise amplifier is tied

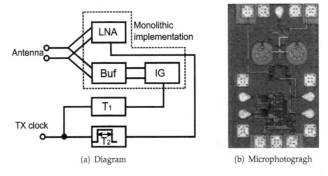

(a) Diagram (b) Microphotogragh

Figure 20. Block diagram and chip photo of the proposed monostatic UWB radar front-end.

together with the output of a buffer amplifier following the impulse generator. An external monoflop and a bandgap reference circuit ensure that the LNA is disabled during the impulse emission. The LNA bias is recovered after the impulse has been transmitted, and it returns to full gain within 1.5 ns. The added parasitics of the buffer are included in the design of LNA.

Fig. 20(b) shows the fabricated IC. In the experimental test, the antenna terminal is connected to two short coaxial cables, each of which feeds into a 10 dB attenuator shorted at the far end. An approximately 1 ns delay is generated by the coaxial cable and attenuator, corresponding to a distance of 30 cm in air. The measured time domain trace at the output of the differential LNA can be seen in Fig. 21. The significant common-mode transients due to the bias switching are completely invisible owing to the balanced setup. The result clearly shows the received impulse echo. Due to a high isolation of the 'cold' low-noise amplifier, the crosstalk from the transmitted pulse is barely visible and will not influence the further processing.

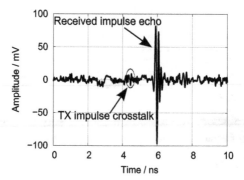

Figure 21. Real time oscilloscope trace showing the functional test of the monostatic radar frontend, displaying transmitted pulse crosstalk and received pulse echo.

3. Signal processing

3.1. Energy detection for UWB communications

For UWB communications, the detection of transmit symbols can be done with a coherent or a non-coherent method. Typically, with perfect synchronization, the coherent detection based on correlation gives a better performance than non-coherent detection. Coherent detection is known to be the optimum method with respect to the bit error rate (BER) for AWGN channels. In case of multipath propagation channels, the transmit impulse and the channel impulse response together (convolved) must be used for correlation detection, and in general more complex signal processing is required. Analog signal processing is considered here because analog to digital conversion for UWB signals is hardly available and requires high power consumption. Unfortunately, the coherent approaches are not well suited for our large bandwidth analog signal processing within the receiving frontend. Energy detection is usually preferred because, if applied in a proper way, no channel impulse response is needed in the receiver. Moreover, energy detection is also robust with respect to synchronization accuracy. In its basic form the energy detector consists of a squaring device, an integrator, a sampler and a decision device.

Pulse position modulation (PPM) and On-off keying (OOK) are modulation techniques that are typically used in combination with energy detection. The data modulation is performed by changing the position or amplitude for PPM and OOK respectively. The detection of PPM is easier to perform since comparing the signal energy at two different intervals is enough, while OOK requires threshold estimation. We look at the BER performance of PPM with respect to synchronization and multipath propagation. Different levels of synchronization in AWGN channels (perfect, ± 20 ps, and ± 40 ps) and a perfect synchronization for a multipath channel are considered. The errors in the synchronization accuracy are uniformly distributed. The transmitted signal is an impulse train of fifth derivative Gaussian functions ($\sigma = 51$ ps) with PPM modulation. The tested channels AWGN channel and a multipath channel with delay spread are assumed to be shorter than 4 ns, which is half of the impulse repetition interval. The correlation receiver uses a template impulse (fifth derivative Gaussian) for correlation, and the correlation is centered at the first strongest path of the channel. The BER for all settings is shown in Fig. 22.

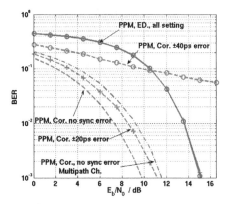

Figure 22. BER performance of PPM modulation for correlator and energy detection with AWGN channel and multipath channel.

The performance of the correlation receiver suffers from synchronization uncertainty and multipath propagation. This is due to the fact that the impulse used in UWB systems is very narrow and the impulse correlation receiver cannot capture all of the signal energy. On the other hand, the energy detector shows good performance also for non-perfect synchronization and multipath channel. We can conclude from the results that, with a performance trade-off, the energy detection is much more robust. Other challenges for implementing energy detection are multiuser capability and interference cancellation. These problems can be solved by the comb filter receiver presented in the following part.

3.2. Comb filter

The received signal power for medical applications are expected to be very small due to high attenuation in human tissue. We propose a receiver based on a comb filter to improve Signal-to-Noise ratio (SNR) before further processing. The comb filter is a feedback loop with an analogue delay and a constant loop gain of one for all frequencies. It is used to perform a coherent combination of the incoming UWB impulses. The feedback loop sums up the number of impulses used for the transmission of a data symbol/measurement and is reset after this. The coherent combination results in SNR improvement, interference suppression which come from different transmitters in a multiuser environment or narrowband interfering signals. Several UWB impulses are transmitted for one data symbol/measurement. One important feature of the concept is that the individual UWB impulses are weighted by +1 or -1 according to a spreading sequence. The UWB transmit signal $s(t)$ can be written as

$$s(t) = \sum_{k=-\infty}^{\infty} \sum_{n=0}^{N-1} c_n p(t - nT_c - kT_s), \qquad (6)$$

where $p(t)$ is a UWB impulse and c_n is the spreading sequence. T_c is the period between two UWB impulses or 'chip period' and $T_s = NT_c$ is symbol period for communications or measurement period for radar/localization application. For communication, assuming a binary transmission, the impulse train of each data symbol can be modulated in different ways. We consider three modulations using the direct sequence spread spectrum technique based on OOK, PPM, Code shift keying (CSK) ,i.e.   DS-OOK, DS-PPM and DS-CSK. For DS-OOK, transmitting a train of impulses represents the data '1', while no impulse signals means '0'. For DS-PPM, the two basic waveforms for a binary transmission are different by time shift. For DS-CSK, the two waveforms result from two different spreading sequences. A decision threshold is not required for DS-PPM and DS-CSK which is a big advantage compared to DS-OOK. The basic waveforms for different modulation techniques are illustrated in Fig. 23.

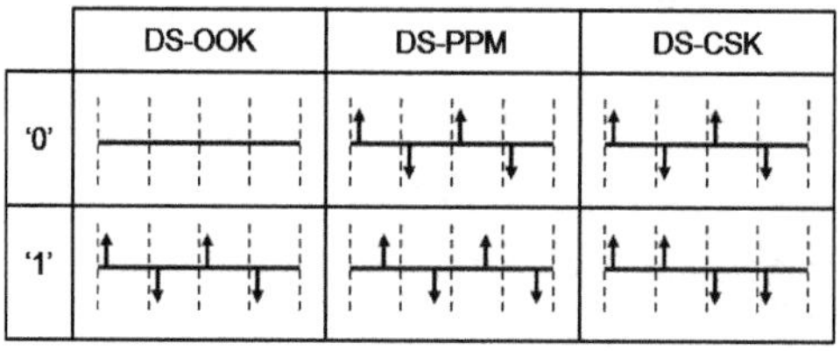

Figure 23. Comparison of UWB transmit signals for different modulation techniques.

Transmitted through the channel, the impulses are affected by the channel impulse response which is expected to be a multipath environment with or without a line-of-sight path. The chip period is chosen to be larger than the multipath spread, with the result that no interchip interference (i.e. no overlapping of channel impulse responses) occurs. We assume the channel to be time invariant within the symbol period. This means that the signal or basic waveform which represents one data symbol/measurement in the received signal consists of a corresponding number of channel impulse responses.

3.2.1. Basic comb filter receiver

The receiver based on the comb filter with remodulation is shown in Fig. 24. The receiver consists of an antenna, a LNA, a multiplier, the comb filter, the energy detector for communications and a correlator for radar/localization applications. This concept allows energy detection in a multipath and multi-sensor environment and amplification of the impulse response in the comb filter delay loop.

In the receiver, the received signal is multiplied with the spreading waveform (i.e. the sequence with rectangular "chips"). If the spreading sequence matches, the result is a periodically repeated channel impulse response with period T_c. After multiplication, the received impulses are delayed and summed up by the comb filter. At the output of the comb filter, only the last T_c period of every T_s is used for further processing which is expected to be an amplified and improved SNR channel impulse response. Interference from different UWB transmitters and other systems are eliminated at the comb filter. The process of remodulation and coherent combination at the comb filter is illustrated in Fig. 25.

Spreading is used not only for the multiuser or multi-sensor purpose, but also for shaping the power spectral density of the transmitted signal. The impulse train may have high spikes in the power density spectrum, because it has a periodic behavior over a longer time interval,

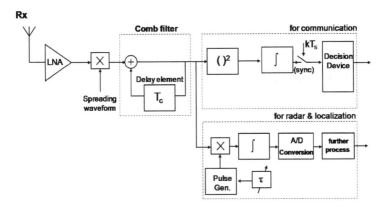

Figure 24. Block diagram of the basic comb filter based receiver.

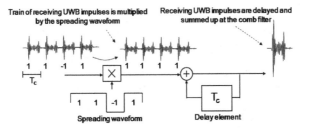

Figure 25. Comb filter principle in the receiver.

and hence it could violate the spectral mask requirement. The longer and more random the spreading sequence, the better. Of course, a direct restriction is given by the desired data rate.

3.2.2. Multiuser interference

In general we will have many transmitters or many sensors and different corresponding spreading sequences. The sequences are selected such that the mutual cross-correlation values are as small as possible. This guarantees that many transmitters can be used at the same time in the same area, e.g. several implants for digital transmission. The signals from all sensors are assumed to be transmitted in parallel with synchronization in a certain time window. The remodulation and the comb filter at the receiver is the crucial part in suppressing the multiuser interference (MUI). The weighting factors of the channel impulse response train after the multiplication are a scrambled version between transmit and receive sequence, if they are not the same. The accumulation at the comb filter would eliminate the MUI and the result of the weight is the cross-correlation values. The illustration of multiplication with the spreading sequence and comb filter accumulation of a signal with two different sequences is shown in Fig. 26.

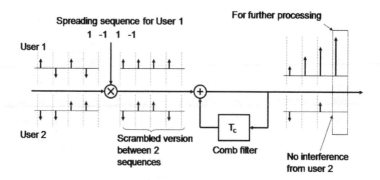

Figure 26. Illustration of multiuser interference suppression at the comb filter.

We assume synchronization in a certain time window between different users. With this assumption, we mainly consider m-sequences because they have low crosscorrelation and are simple to generate. Alternatively, binary zero-correlation-zone (ZCZ) sequence can also be considered. Unlike m-sequences, these sequences have a crosscorrelation of zero within a small window but the drawback is a smaller number of sequence in one set [40]. Binary ZCZ sequence sets were investigated in [37] for communication because of their potential to completely eliminate multi-sensor and interchip interference (ICI). A near-far problem was taken into account and we can see an advantage of this sequence set.

3.2.3. Signal-to-noise ratio improvement

The SNR of received signals can be significantly improved after the coherent combination in the comb filter, since the signal power grows quadratically, while the noise power grows only linearly. We first consider ideal components where there is no distortion in delay line and the comb filter has no loop gain. For the investigation of the SNR improvement, an AWGN channel is sufficient. A train of modulated UWB impulses and additive noise is the input signal to the multiplier. The SNR at the input of the multiplier is compared to the SNR after the comb filter to calculate the comb filter processing gain G_p. Fig. 27(a) shows an example of an input signal with a SNR = -15 dB (upper) compared to the output signal (lower). The SNR improvement can be seen clearly as the receiving impulse becomes visible after a few iterations. This comb filter signal processing results in a SNR improvement of $10 \log(N)$ dB, where N is the number of iterations. The UWB signal is algebraically added, therefore the signal energy is increased by a factor of N^2. On the other hand, the noise contributions in each chip are added in power, and therefore the noise energy within the symbol interval is increased only by a factor of N. For communications, the BER of PPM with correlation and energy detection also show the same improvement. The BER performance for DS-PPM with $N = 1$ and $N = 63$ using M sequence is shown in Fig. 27(b) . Using several chips per symbol with the comb filter approach gives a performance gain with respect to SNR but not to E_b / N_0 due to the difference in the data rate. For UWB transmission, the SNR cannot be improved by increasing the transmitted energy because of the spectral mask limitation. A trade-off between data rate and the SNR improvement at the receiving side has to be made.

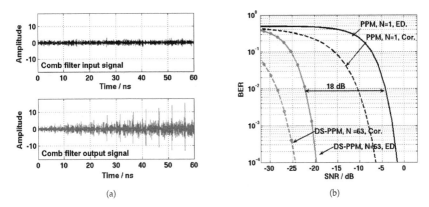

Figure 27. (a) Input and output of the comb filter with remodulation. (b) BER performance of DS-PPM with N = 1, 63.

For physical systems this calculation is only valid if no instability is introduced. Therefore the values of the gain G_c of the comb filter can only be in the range of 0-1, because otherwise an oscillation occurs. G_c is designed to be one but considering real components, a loss in the loop is possible. The comb filter processing gain G_p is a function of the gain G_c of the comb filter and the number of iterations N. It can be calculated as

$$G_p = 10 \cdot \log_{10}\left(\frac{\left(\sum_{n=0}^{N-1} G_c^n\right)^2}{\sum_{n=0}^{N-1} G_c^{2n}}\right) \quad (7)$$

The maximum processing gain G_p of the comb filter is 10 log(N) dB, when $G_c = 1$. The processing gain is reduced if G_c is less than one. The relations between G_p, G_c and N are shown in Fig. 28. We can see from Fig. 28(a) that the processing gain G_p saturated in the case where G_c is less than one because the impulse energy vanishes after some iterations. With a higher number of impulses per symbol, the comb filter loop gain has to be controlled more precisely as shown in Fig. 28(b).

3.2.4. Narrowband interference

Since UWB covers a very large bandwidth, strong interference within the band is possible and can cause problems at the receiver. Comb filter in combination with multiplication with the spreading waveform can suppress narrowband interference very well. Only a periodic signal that has a period which equals one or multiples of the comb filter delay can go through the comb filter. The comb filter transfer function $H(f)$ is given as follows:

$$H(f) = \sum_{k=-\infty}^{\infty} T_s \cdot \text{sinc}(T_s(f - k/T_c)) \exp(-j\pi f T_s) \quad (8)$$

We can see that the transfer function of the comb filter consists of several peaks. The peaks could be seen as tunnels that allow only signals with specific frequencies to pass. After the

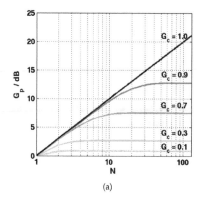
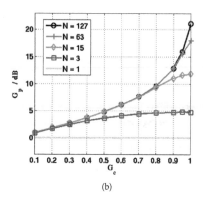

(a)   (b)

Figure 28. Relationship between processing gain G_p, comb filter loop gain G_c and impulses per symbol N.

multiplication with the spreading waveform in the receiver, the spectrum of the UWB signal has a shape matched to the transfer function of the comb filter (signal with period T_c). On the other hand, the narrowband interferer is spread by this multiplication and only some frequency components could go through the comb filter. The width of each peak depends on the number of chips per symbol. It gets smaller as the number of summation steps in the comb filter increases and as a result more interference is suppressed.

The improvement of the signal-to-interference ratio (SIR) is demonstrated in Fig. 29(a). The signals at the input of the comb filter are the UWB signal and a narrowband interference consisting of an IEEE 802.11a OFDM WLAN signal with a bandwidth of 16.66 MHz and a center frequency of 5.2 GHz. In this example, the input SIR of -15 dB is improved by 10 dB for $N = 63$. In addition, the BER performance of DS-OOK and DS-PPM in an AWGN channel with the same interference is shown in Fig. 29(b). The performance for both methods is improved with increasing number of chips per symbol. The degradation of the performance due to the narrowband interference for DS-PPM is much less than that for DS-OOK. The narrowband interferer gives a contribution to both integrator outputs for DS-PPM, and because the outputs are compared, the influence is reduced. For the DS-OOK the influence remains.

3.2.5. Shortened delay comb filter

The main challenge for implementing the comb filter based receiver is to realize the true wideband analog delay element. Shortening the delay means that overlapping channel impulse responses can occur at the receiver, the channel impulse response becomes longer than the chip interval T_c. As a result, after the comb filtering we get an amplified window-cut-out of the true impulse response with the window width T_c. It is shown in [35] that we can control the position of this window by adjusting the spreading sequence at the remodulation. The property of being able to extract different parts of the channel impulse response gives an opportunity to construct a receiver by using a rake concept. This means that for each part of the impulse response we have one rake branch where we calculate the energy. The modulation technique that is well-suited for this structure is DS-CSK. The block diagram

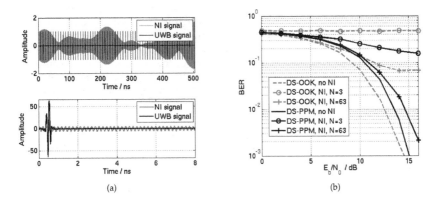

(a)   (b)

Figure 29. (a) Input and output of the comb filter with remodulation. (b) BER performance with SIR = -15 dB.

of the receiver for this concept is shown in Fig. 30. Note that the spreading sequences for the rake branches are cyclic-shifted versions of one single sequence. For DS-CSK, two parallel branches with different corresponding spreading sequence sets are needed. The results from two branches are then compared.

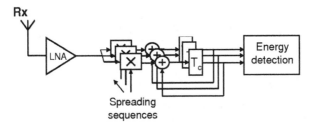

Figure 30. Block diagram of a rake-like comb filter receiver with energy detection.

In comparison to our basic concept described before, the delay is shorter. If the delay is shortened, for example, from 12 ns to 2 ns, this is much more realistic with regard to the realization. The performance of the new concept was verified by simulations. We used UWB impulses within a band from 4 GHz to 6 GHz. Note that this band is due to the Wireless Body Area Networks (WBANs) channel model we used [12] but it is not important for our concept, it could be any. The channel model consists of two groups of paths with a lognormal fading statistic for each group. We used 4000 channel realizations. The total propagation time of the channels is 12 ns. Eight users/sensors with the same average receive energy are considered. The spreading sequences were m-sequences with $N = 127$. The receive signal at the input of the comb filter is shown in the upper part of Fig. 31(a). Additive noise is not considered here to focus only on the ICI and MUI suppression. We can see that both, ICI and MUI are very strong here. The SIR in this example is -17 dB. In the lower part of Fig. 31(a) the original channel impulse response (black) is compared to the results taken from different rake branches (red). From the results we can see that the interference is suppressed very well.

We also look at the BER performance of this new concept in Fig. 31. A WBAN channel is used. For our original basic comb filter concept, $T_c = 12$ ns and 6 ns are considered. We label them as 'Basic 1' and 'Basic 2', respectively. For the rake-like receiver, we consider $T_c = 2$ ns with 6 rake branches which yield equivalent integration time of 12 ns. We can see that the performance of our original concept is getting worse if the delay is shortened. There is a loss of about 4 dB at the BER for 10^{-4}. The result for 'Rake-like' is comparable to 'Basic 1'. The results show that we can achieve a very similar performance to the original concept with much shorter analog delay elements.

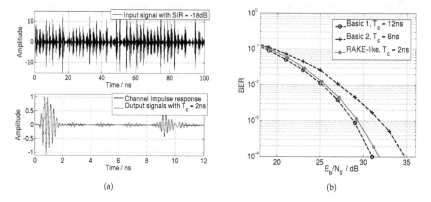

(a)   (b)

Figure 31. (a) Input signal of the comb filter (upper). Comparison between the actual channel impulse response and the output signal of the rake branches (lower). (b) Comparison of the bit error rate from different receivers.

3.3. Particle filtering

In this part we want to address an algorithm for movement tracking which can be used for radar and imaging application. The background is the application of IR-UWB for biomedical diagnostics, e.g. vital sign detection (like breathing/heart rate) and also tracking of body movements to compensate for errors which degrade the quality of inner body imaging. Simple methods such as tracking the maximum/minimum of the receive signal in a specific area perform reasonably well. A problem occurs if the peak of the signal cannot be easily identified. In cases with high attenuation, the estimation can become very bad for those simple methods. Another problem is caused by stationary echoes (clutter). Particle filtering can help in this situation. It is a technique which implements a sequential Bayesian estimation by using Monte-Carlo methods. Particle filtering is commonly known to be used in localization applications and tracking in dynamic scenarios [18]. The estimation uses a movement model to incorporate the temporal correlation of the change of unknown parameters.

The Bayesian estimator finds unknown parameters x_k (signal delays) from a set of measurement signals $z_{1:k}$ using a posterior density function $p(x_k|z_{1:k})$. It is very common that the algorithm is processed in a recursive manner. In summary, the algorithm consists of a 'prediction' phase and an 'update' phase, where the prior density $p(x_k|z_{1:k-1})$ and the posterior density $p(x_k|z_{1:k})$ are estimated every time step k.

Prediction:

$$p(\boldsymbol{x}_k|\boldsymbol{z}_{1:k-1}) = \int p(\boldsymbol{x}_k|\boldsymbol{x}_{k-1})p(\boldsymbol{x}_{k-1}|\boldsymbol{z}_{k-1})d\boldsymbol{x}_{k-1} \qquad (9)$$

Update:

$$p(\boldsymbol{x}_k|\boldsymbol{z}_{1:k}) = \frac{p(\boldsymbol{z}_k|\boldsymbol{x}_k)p(\boldsymbol{x}_k|\boldsymbol{z}_{1:k-1})}{p(\boldsymbol{z}_{1:k}|\boldsymbol{z}_{1:k-1})} \qquad (10)$$

We can see that the likelihood function $p(\boldsymbol{z}_k|\boldsymbol{x}_k)$ and the movement model $p(\boldsymbol{x}_k|\boldsymbol{x}_{k-1})$ are important items in Bayesian estimation. Unfortunately the posterior density is usually intractable but there are several ways to implement the algorithm. Particle filtering deals with this problem by using samples (particles) with associated weights to represent the posterior density. If the number of particles is sufficiently large, the estimates reach optimal Bayesian estimation.

A setup consisting of a bistatic UWB transceiver and a moving metal plate is used for demonstrating the use of particle filtering in IR-UWB tracking. A 5th derivative Gaussian impulse fitting to the FCC mask is used and the repetition rate is 200 MHz (i.e. $T_s = 5$ ns). The corresponding distance is 75 cm. The metal plate moves periodically within 5 mm range. A schematic block diagram of the setup is shown in Fig. 32(a). The receive signal consists of two main contributions which can be seen in each period, corresponding to two-path propagation on the channel. The first path is the direct path between the transmit and the receive antennas (direct coupling signal). This signal has very strong visible ringing due to signal reflection from the impedance mismatch. The second signal is a signal reflected from the moving metal plate. An example of a receiving signal with a target 30 cm away is shown in Fig. 32(b). More details about the setup can be found in Sec. 4.1.

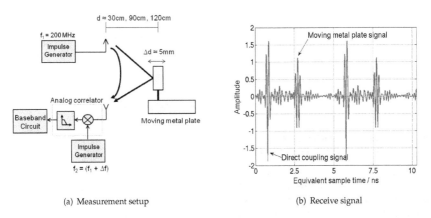

(a) Measurement setup   (b) Receive signal

Figure 32. Measurement setup for target movement tracking and example of a periodic receive signal after the correlator.

Particle filtering with a 2-path model is used for tracking the transmission delay of the two paths. The signal after the impulse correlation can be represented in discrete-time in each period and is considered as measurement signal z_k. The delay of each paths form the state vector x_k. More details on the setup can be found in [36].

In the following, tracking results for a moving metal plate with distances of approximately 90 cm and 120 cm are discussed. Since the distance of ambiguity for the measurement setup is 75 cm, we use the knowledge that the target is in the 75-150 cm range. This is sufficiently large for our target application. The reflected signal appears one period after the original impulse was transmitted. Particle filtering with 1000 particles is considered and the results are compared with a conventional maximum tracking method. The movement of the first path (direct coupling) and the second path (metal plate) are tracked simultaneously.

We first consider a setup with the moving target at approximately 120 cm. An example for one period of the receive signal is shown in Fig. 33(a). The reflected signal is located at around 8.8 ns. In this example, the reflected signal can be easily recognized. Fig. 33(b). shows the tracking results of the moving metal plate from both methods (particle filtering and maximum tracking for comparison). The tracking results fit well with each other and the small movement of 5 mm was estimated correctly. We can see that the particle filtering is more robust. This improvement comes from the fact that the movement model incorporates the temporal correlation of the change of the channel delays in different time steps. The conventional method does not use this information and the results can change rapidly. The particle filter needs some iterations to converge to the correct estimate.

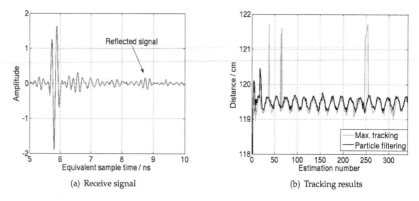

(a) Receive signal   (b) Tracking results

Figure 33. Receive signal with target distance ≈ 120 cm and tracking result for target distance ≈ 120 cm from conventional maximum tracking and particle filtering.

In this part, we consider the case where the noise is strong and the peak of the signal cannot be easily identified. To reduce the SNR of the signal, Gaussian noise was added to the measurement data of the previous part, so that the SNR was approximately 0dB. An example for one period of the receive signal z_k is shown in Fig. 34(a). We can see that the target signal is not clearly visible anymore and the peak value is disturbed strongly by noise. The tracking results are shown in Fig. 34(b). The conventional method does not work. The particle filtering still gives good estimates, because it does not consider only the maximum point in the target signal but the waveform as a whole. The movement model also plays a role in this improvement.

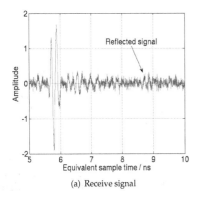

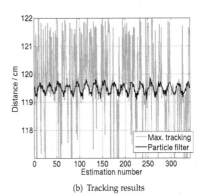

(a) Receive signal   (b) Tracking results

Figure 34. Receive signal with target distance $\approx 120\,\text{cm}$ with additional noise and tracking result for target signal $\approx 120\,\text{cm}$ with additional noise from conventional maximum tracking and particle filtering.

Now we consider the tracking of the reflected signal for the moving metal plate at a distance of 90 cm. An example of a receive signal is shown in Fig. 35(a). We can see that the target signal is in the same interval as the ringing of the direct coupling signal. The amplitudes of the target signal and the ringing are comparable. It is not easy to distinguish between these two signals anymore. This is the same situation as in radar where we have cluttering. It can cause a bias to the estimation. A comparison of the tracking results from maximum tracking and particle filtering is shown in Fig. 35(b). We can see that the maximum tracking performs very badly because of the bias. The tracking results from particle filtering are very good. The use of the multipath propagation model eliminates the error bias given by the clutter (direct coupling).

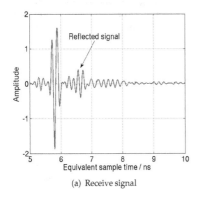

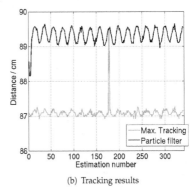

(a) Receive signal   (b) Tracking results

Figure 35. Receive signal with target distance $\approx 90\,\text{cm}$ and tracking result for target distance $\approx 90\,\text{cm}$ from conventional maximum tracking and particle filtering.

Usually, the drawback of the particle filter is its complexity. In the applications considered here, this drawback is not so serious, because our system model is relatively simple. Using particle filtering in parallel with conventional methods and exchange information between both methods can also help to reduce complexity.

3.4. Surface estimation and subsurface localization algorithms

In this section we present algorithms which can be applied for the localization of actively transmitting beacons inside of the human body. The targeted application is the tracking of catheters equiped with UWB transmitters. In this context, the use of active transmitters would mitigate the challenges related to the high attenuation of electromagnetic waves in human tissue, which makes purely passive localization extremely difficult [3].

A similar approach has been investigated in the field of ultrasonics, where catheter-mounted ultrasound transducers in combination with external arrays of imaging transducers are used to track the catheter position [25]. The advantage of UWB catheter localization is a contactless measurement setup with receivers placed in air around the patient while ultrasound transducers have to be placed directly on the body.

In subsurface imaging and localization problems, where sensors are not directly in contact with the medium, the permittivity contrast between air and the medium cannot be neglected as it leads to a different wave propagation velocity inside of the material. In case of medical applications, a relative permittivity of human tissue between 30 and 50 in the FCC UWB frequency range has to be considered [13]. UWB signals are therefore strongly reflected at the air-to-body interface. These reflections are beneficial to surface estimation applications using UWB pulse radars. For in-body localization, however, they constrain a signal emission from inside of the body. In order to overcome the strong reflection losses, we therefore propose a system that combines an active transmitter inside of the body with an array of radar transceivers outside of the body. The sensor array acts as both a surface scanner and a receiver recording the time of arrival (ToA) of a signal transmitted from inside of the body. A determination of the exact body shape prior to localization is necessary to cope with refraction effects at its surface and the change of wave propagation speed from air to tissue. This distiguishes the proposed in-body localization method from most through-dielectric localization problems like through-the-wall imaging where a plane boundary between air and the target medium is assumed [1].

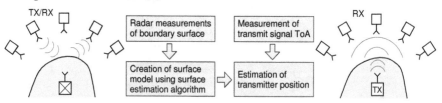

Figure 36. Proposed localization procedure of a transmitter buried in a dielectric medium. In the first step the sensor array is used to scan the surface (left), in the second step it only receives the signal transmitted from inside of the medium (right).

Fig. 36 summarizes the proposed localization approach of a transmitter placed in a dielectric medium. In a first step, the array of radar sensors is used to measure the distance to the boundary surface. These measured distances together with the known antenna positions are the input of the surface estimation algorithm which returns a model of the boundary surface.

In the second step, the transmitter inside of the dielectric is switched on and the radar sensors operate in receive mode recording the time of flight of the transmitted signal. Finally, we analyze all the acquired data and determine the position of the transmitter.

Several methods to estimate the surface of a highly reflective medium using UWB pulse radar sensors have been investigated in recent years [17, 29]. Some of these imaging algorithms, however, need extensive preprocessing of the measurement data or suffer from high complexity and computation time. In the first part of this section, we derive a simple and easy to implement 3D surface estimation algorithm based on trilateration. In the second part, building on this surface estimation method, we present an approach for the localization of transmitters inside an arbitrarily shaped dielectric medium taking into account its surface profile.

3.4.1. Surface estimation algorithm based on trilateration

Radar measurements with quasi-omnidirectional antennas only provide information about the target distance, but not about its direction. This makes surface imaging an inverse problem which can only be solved by combining measurement results from different antenna positions. In this context, target ranging using trilateration means determining the intersections of spheres, the radii of which correspond to measured target distances. The underlying assumption for using trilateration as a surface estimation method is that two neighboring antennas are "seeing" the same scattering center. As with other imaging algorithms this assumption can lead to inaccuracies of estimated target points.

The imaging principle shall first be explained using a two-dimensional example. Fig. 37(a) shows the measurement scenario of a linear array of monostatic radar transceivers arranged along the x-axis scanning the surface of a target in z-direction. Each array element measures the distance to the closest point on the target. Two exemplary measurement points X_n and X_{n+1} are picked out, and semi circles with radii corresponding to the measured target distances are plotted around the antennas. The estimated surface point is the intersection of the two circles.

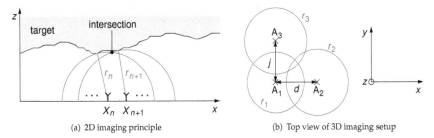

(a) 2D imaging principle   (b) Top view of 3D imaging setup

Figure 37. Cross section of a 2D imaging problem using a linear array of monostatic radar transceivers along the x-axis (a) and top view of a 3D imaging setup showing three measuring points of an antenna array in the x-y-plane (b).

Three-dimensional imaging demands for a third antenna position located in a different dimension. This setup is shown as a top view in Fig. 37(b). At each of the three positions a target distance r_i is measured which leads to a set of spheres with radius r_i around the respective antenna position A_i, as defined by the system of equations

$$r_1^2 = x^2 + y^2 + z^2, \tag{11}$$

$$r_2^2 = (x - d)^2 + y^2 + z^2, \tag{12}$$

$$r_3^2 = x^2 + (y - j)^2 + z^2, \tag{13}$$

where d and j are the distances between two antennas in x- and y-direction, respectively. For simplicity, the first antenna position A_1 shall be at the center of the coordinate system. The above equation system is valid for a planar antenna array. In case of a curved antenna array an offset z-value has to be inserted.

The initial assumption that all three antennas are "seeing" the same target has to be assured by comparing the measured target distances $r_{1\text{-}3}$. If the difference between these distances is small enough, the assumption can be considered valid. For the above equations this precondition can be formulated as follows:

$$|r_1 - r_2| \leq T_{s,x} \quad \text{and} \quad |r_1 - r_3| \leq T_{s,y} \tag{14}$$

A threshold T_s in the range of about half the antenna distance has shown good results.

If the conditions in (14) are fulfilled, the target surface point of interest can be calculated by intersecting the three spheres. The coordinates (x, y, z) of the intersection are

$$x = \frac{r_1^2 - r_2^2 + d^2}{2d}, \tag{15}$$

$$y = \frac{r_1^2 - r_3^2 + j^2}{2j}, \tag{16}$$

$$z = \pm\sqrt{r_1^2 - x^2 - y^2}. \tag{17}$$

These coordinates are offsets refering to the position of the first antenna A_1. The sign in (17) depends on the arrangement of the radar transceivers. Here, we assume that the antennas are oriented towards positive z-values.

The necessary steps of the presented trilateration algorithm can be summarized as follows:

1. Pick three neighboring measurement points in two different dimensions (here: along the x- and y-axis).

2. Extract the target distance from the recorded radar measurement data at each antenna position. Multiple target responses per measurement are possible.

3. Check if the differences between the measured distances satisfy the trilateration condition in eq. (14).

4. If the previous condition is fulfilled, calculate the target coordinates using eq. (15)-(17).

5. Repeat the two previous steps if higher order reflections exist, or otherwise start over with the next three measurement positions.

3.4.2. Subsurface localization algorithm

The given parameters of the localization problem are the shape of the dielectric medium containing the transmitter, its distance to the antenna array and the ToA of the localization signal at each array element. Since all the wave propagation effects are reciprocal, our problem can also be regarded in a reverse way: At each receiver position the transmission of a short pulse with a delay corresponding to the respective previously measured ToA is assumed. In order to get the original beacon position we have to find the spot where all these virtual pulses would superimpose, i.e. the intersection of the impulse wavefronts inside of the medium [27].

According to the Huygens-Fresnel principle a refracted wavefront can be represented by an infinite number of spherical waves which originate from points on the boundary surface reached by the incoming wave. This is shown in Fig. 38(a) for a pulse transmitted from an antenna at position (3,0) towards a dielectric half space. In this 2D example, 20 source points of radial waves on the dielectric surface are considered. The radii $r_{i,n}$ of the secondary waves are calculated from the measured ToA at the receiving antenna and the length of the ray $p_{i,n}$ between the respective antenna position n and surface point i:

$$r_{i,n} = \frac{1}{\sqrt{\varepsilon_r}} \left(c_0 \cdot \text{ToA}_n - p_{i,n} \right), \tag{18}$$

where c_0 is the speed of light. The division by $\sqrt{\varepsilon_r}$ accounts for the different wave propagation speed in the dielectric medium. The envelope of all radial waves corresponds to the wavefront we are looking for. By repeating the procedure for every antenna element of the receiver array we get a set of wavefronts as illustrated in Fig. 38(b). Finally, we determine the intersection of these wavefronts to obtain the transmitter position inside of the medium. A 3D localization problem is solved in an analog way with envelopes of spheres leading to intersecting 3D wavefronts. Here, however, the derivation of the localization procedure shall be limited to the 2D case because of simpler graphical representations.

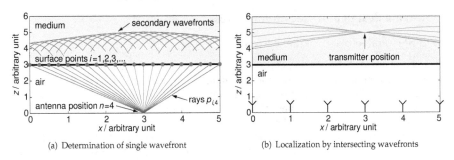

(a) Determination of single wavefront   (b) Localization by intersecting wavefronts

Figure 38. Determination of the 2D wavefront shape inside a medium with $\varepsilon_r = 30$ for a signal transmitted in air at position (3,0) by evaluating the envelope of radial secondary wavefronts (a). Intersections of wavefronts corresponding to six different antenna positions indicate the transmitter location at (3,5) (b).

The wavefront shapes in Fig. 38(b) agree with the hyperbolic approximation of wavefronts in dielectric half spaces [28]. With more complex boundaries, however, the analytical calculation of refracted wavefronts is no longer practical, while the approach presented here is independent of the surface shape. An example of a transmitter placed behind a more complex surface is presented in Fig. 39. The times of flight between the transmitter at (5 cm,5 cm) and the individual elements of the receiver array at $z = -20$ cm are calculated using electromagnetic field simulation software [6].

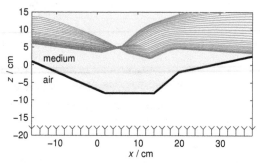

Figure 39. Calculated wavefronts in a medium with ε_r=10 based on EM field simulations. The receiving antennas are placed along the x-axis at z=-20, the transmitter is positioned at (5 cm,5 cm).

The above mentioned 2D localization procedure leads to a belt of refracted wavefronts with a focussing point where the receiver has been placed in the simulation. To decrease calculation time it is also possible to search for the narrowest point in the wavefront belt instead of calculating the intersections. This leads to an estimated transmitter position of (4.60 cm,4.94 cm), having an error of about 4 mm.

It is obvious that in the example of Fig. 39 a smaller number of wavefronts and thus less receiving antennas would be sufficient for a successful localization of the transmitter. But in practical applications this high number of antennas might still be required as a dense sensor array is rather needed for surface estimation than for solving the localization problem.

4. Systems design and measurement results

4.1. Single-ended bistatic radar system

At first a single-ended radar demonstrator was developed in the project, with which the potential of impulse-radio UWB sensing is evaluated. The demonstrator is built combining commercially available components for the low frequency operation control and components tailored for UWB operation dealing with the signals in the 3.1-10.6 GHz band. The UWB components are developed and fabricated in the aforementioned SiGe HBT production technology. A block diagram of the single-ended bistatic radar system is depicted in Fig. 40. The sensor uses separate antennas for transmitter and receiver to avoid losses due to power divider structures on the feedline of a single transceiver antenna. Besides a heavy crosstalk into the low-noise amplifier (LNA) of the receiver is avoided by using two antennas.

An ultra-broadband Vivaldi antenna is chosen, which consists of an exponentially tapered slot on a microstrip substrate. The transition from microstrip to slot line is done by a Marchand balun, as discussed in [22]. On the feeding line of the transmit antenna an impulse generator IC is mounted, which emits an impulse with a shape very similar to the fifth derivative of a Gaussian bell shape with a standard deviation $\sigma = 51$ ps. This impulse shape fits into

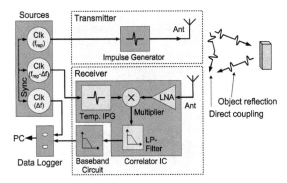

Figure 40. Block diagram of the single-ended bistatic radar system.

the 3.1-10.6 GHz UWB spectral mask allocated by the FCC in the United States and has a voltage amplitude of 600 mV peak-to-peak [8]. The impulse generator radiates an impulse at every rising slope of an input trigger signal. Here a sinusoidal signal is used to trigger the impulse generator, so at every rising edge of the sinusoidal signal an impulse is emitted which results in a continuous impulse train. The sinusoidal signal is supplied from one source of the direct-digital-synthesizer (DDS) AD9959. All four clock sources of the AD9959 (three are used) are synchronized among each other to allow a phase and frequency stable operation between the signals. The transmitter is adjusted to generate an impulse train with a repetition rate of $f_{rep} = 200$ MHz. To reduce impulse-to-impulse jitter, spurious emissions of the 200 MHz sinusoidal trigger signal are filtered by a narrowband helix filter. The generated impulse train is continuously radiated by the antennas, is reflected at the desired object and enters the receiver. The reflection at the object causes a phase inversion to the impulses, therefore it is received with inverted amplitude. Additionally the impulse train is fed to the receiver by a direct and non-inversed coupling between the antennas.

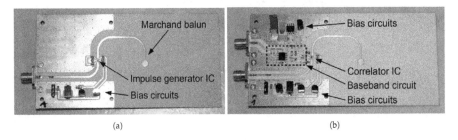

(a)   (b)

Figure 41. Picture of (a) Vivaldi transmit antenna with mounted impulse generator IC and (b) Vivaldi receive antenna with correlator IC and baseband circuit.

These signals are processed in the receiver by a monolithic correlator IC, which consist of a UWB LNA, a four-quadrant multiplier, a template impulse generator generating a fifth Gaussian derivative impulse corresponding to the transmit impulse, and a first integrating low-pass filter with a cut-off frequency of 800 MHz [9]. The template impulse generator is driven by a second clock source of the AD9959 at a repetition rate of

$f_{rep} - \Delta f = 200\,\text{MHz} - 25\,\text{Hz}$. This sinusoidal clock signal is filtered as well with a helix filter to improve jitter performance. The output signal of the correlator IC is processed with a baseband circuit, where it is amplified and further integrated by a low-pass filter with a cut-off frequency of 25 kHz. After this it is fed to a data-logger, which samples the generated signals with a sampling period of 30 μs and transfers them to a PC for further processing. Additionally a synchronizing signal of $\Delta f = 25\,\text{Hz}$ from the third DDS clock signal is sampled

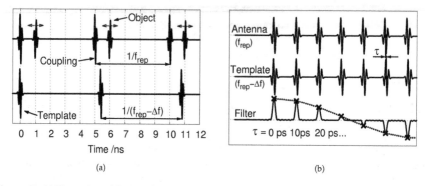

(a)   (b)

Figure 42. (a) Illustration of the template impulse sweeping through the antenna receive signal. (b) Illustration of the evolving correlation signal from receive and template signal in the region where the impulses overlap.

and transferred to the PC as well. In the PC a post-processing of the continuous data stream is done using a custom software written in Lab Windows/CVI. Pictures of transmit and receive antenna with mounted components are shown in Fig. 41. A detailed description and supporting measurements of the hardware can be found in [7, 31].

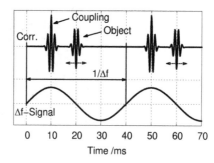

Figure 43. Illustration of the correlation signal at the receiver antenna output port, together with the Δf-signal for separation of the repetition.

To determine the distance between sensor and desired object and the variation of the distance between them, a time-of-flight measurement is applied using a sweeping impulse correlation as explained in [10]. The sweeping impulse correlation is very similar to an undersampling technique and avoids high-speed sampling of the gigahertz-range UWB impulses. This method is illustrated in Fig. 42 by showing the signals appearing in the receiver. In the upper trace of Fig. 42(a) the received signal at the output of the LNA in the correlator IC is shown. An impulse from both, the non-inverted direct coupling and the inverted reflection at the object is shown. The impulses are repeated at the repetition rate f_{rep}. The template impulse in the correlator IC, shown in the lower part of Fig. 42(a), is operating at a repetition rate $f_{rep} - \Delta f$. Therefore it is continuously changing its time alignment to the received signals and appears sweeping through the impulse sequence.

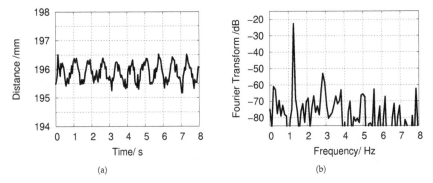

Figure 44. Measurement of sinusoidally moving metal plate placed in front of the antennas at a distance of 19.6 cm and a deviation of approx. 1 mm in (a) time and (b) frequency domain.

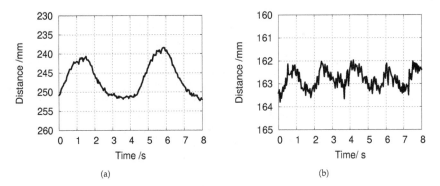

Figure 45. Time domain breathing measurements of (a) a male test person lying on the back and (b) a seven-week old child sleeping in bed.

At each position of the impulses the cross-correlation is computed by multiplying the signals and integrating them. This operation can be seen in Fig. 42(b) in the region where one received impulse and the template impulse are overlapping each other. The lower part of the figure illustrates the output of both filters. The solid line represents the output signal of the 800 MHz filter which integrates each single impulse. The dashed line shows the output signal of the baseband filter with a cut-off frequency of 25 kHz, which integrates a multitude of single impulses. By this procedure, a cross-correlation curve of two fifth Gaussian derivatives is developing. The correlated curve of the above illustration example is depicted in Fig. 43. The correlation impulses from the direct coupling and the reflection at the object can be distinguished. As discussed, the reflection at the object is mirrored, because the impulse is inverted by the reflection. This signal is present at the output of the receive antenna and is sampled by the data-logger. The correlation signal is continuously repeated with a repetition rate of Δf. For a separation of the correlation sweeps, the Δf-signal is sampled as well by the data-logger. When the object under investigation is now moving, the part of the correlation signal coming from the object reflection is correspondingly changing its alignment to the Δf-signal and the movement can be measured. The movement determination of the object is continuously done by software in the PC. First a separation of the correlation sweeps by the rising slope of the Δf-signal is performed. Then both slopes of the correlation curves, the slope from the object and the slope from the direct coupling, are tracked and their positions continuously monitored. Tracking both slopes yielded best precision, compared to tracking the minimum of the correlation curve or only the slope of the correlation signal from the object [31].

To measure the precision of the sensor a metal plate is placed in front, which is mounted on a sledge driven by an eccentric disk, moving the metal plate forward and backward with an approximately sinusoidal deviation. Fig. 44(a) shows a time domain record of a movement measurement with the metal plate placed at a distance of 19.6 cm, a deviation amplitude of approximately 1 mm and a repetition rate of around 1.35 cycles/s. The movement is clearly resolved by the measurement. In Fig. 44(b) the calculated spectrum of the measurement can be seen. The frequency maximum is very clearly visible and verifies a precision of the demonstrator in the millimeter to sub-millimeter range.

In a further measurement the sensor is pointed to the abdomen of a male test person lying on the back at a distance of approximately 25 cm[1]. At the abdomen the largest breathing amplitude occurs. Fig. 45(a) shows a breathing measurement in case the person is breathing normally. The breathing amplitude exceeds 10 mm and the repetition rate is around 2.5 cycles/s. For a further measurement the demonstrator is placed towards a sleeping seven-week old child lying on the back at a distance of approximately 16.3 cm. Fig. 45(b) shows a rhythmic breathing period with a movement of around 1 mm in the direction of the sensor and a repetition rate of 1.5 cycles/s. These measurements show, that the sensor can be used to monitor the breathing of adults and infants lying on the back and that breathing patterns can clearly be detected using the single-ended bistatic impulse-radio UWB radar demonstrator.

4.2. Differential bistatic radar system

A differential bistatic radar system for detecting vital signs was also developed. It follows the approach described in Sec. 4.1, but with fully differential ICs as described in Sec. 2.2 and Sec. 2.3.3 and with significantly reduced power consumption. Here two dipole-fed circular slot antennas are chosen instead of the the Vivaldi antennas. Both the differential impulse

---

[1] Measurements on humans using the single-ended bistatic radar sensor have been approved by the ethic commission of Ulm University.

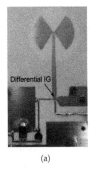 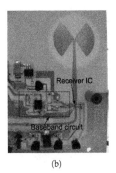

(a)   (b)

Figure 46. Photographs of transmit antenna with mounted differential impulse generator IC and the complete receiver with RF frontend IC and baseband circuit.

generator and the correlation receiver front-end are mounted chip-on-board at the feeding points of the dipole antennas. Fig. 46 shows the pictures of transmit and receive antennas with mounted differential ICs.

Using the same DDS clock generator and post-processing software as described in Sec. 4.1, the ability of the realized differential bistatic radar system for tracking a metal plate which moves back and forth with a sinusoidal deviation is demonstrated. The measured result in the time domain can be seen in Fig. 47(a). A deviation amplitude of around 1.5 mm can be

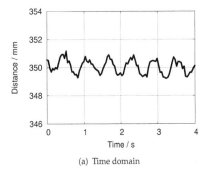 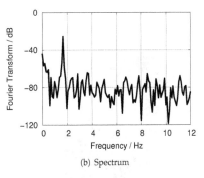

(a) Time domain   (b) Spectrum

Figure 47. Measurement results of a moving metal plate in front of the bistatic radar with a distance of 35 cm in time and frequency domains.

clearly seen. Fig. 47(b) shows the calculated spectrum information of the measurement. The maximum point is clearly visible and indicates the movement frequency of the metal plate.

A common application for this radar system is the detection of vital signs. Here, an adult male with pronounced tachypnea is seated 5 cm from the radar. Fig. 48(a) shows the recorded time domain data. The breathing pattern is clearly visible and its amplitude is around 5 mm. The time domain data is Fourier transformed to frequency domain as shown in Fig. 48(b). It clearly indicates that the respiration rate is around 35/min.

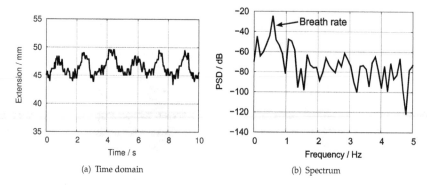

(a) Time domain   (b) Spectrum

Figure 48. Time domain and frequency domain measurements of vital signs of a male test person standing in front of the bistatic radar.

In a further measurement, surface estimation of a container filled with a sugar solution whose properties are similar to those of human tissue was performed by moving the radar up and down in 2 cm steps along both the x- and y-axis. For demonstrating that the radar system is capable for this application, only the lower part of the container is scanned. The trilateration-based imaging algorithm derived in Sec. 3.4 is applied. The photograph of the container is shown in Fig. 49(a), and a cloud of estimated surface points representing the front of the target can be seen in in Fig. 49(b). The result clearly indicates the distance from the radar sensor to the container and the planar surface structure.

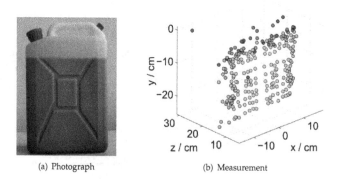

(a) Photograph   (b) Measurement

Figure 49. Photograph of the liquid container and measured cloud of estimated surface points.

4.3. Communcation with implants

In this section we address another application of UWB technology in medicine, the communication with implants. Impulse-based UWB technology is a promising solution for future implanted medical devices demanding data rates in the Mbit/s range and a low power consumption. Here, we present a demonstration system for uni-directional data transmission

between a transmitter inside tissue-mimicking liquid and a receiver placed inside or outside
the medium. These two measurement scenarios are illustrated in Fig. 50.

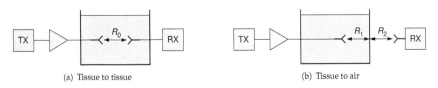

(a) Tissue to tissue   (b) Tissue to air

Figure 50. Measurement scenarios for the demonstration of impulse-based data transmission for
implants using an energy detector. An amplifier is applied on the transmit side to cope with the high
attenuation of the phantom liquid.

As discussed in Sec. 3.1, many different approaches for IR-UWB receivers are feasible.
Non-coherent detection is chosen here because of the challenges regarding synchronization
in coherent receiver setups and because of the dispersive properties of human tissue, which
lead to unpredictable shapes of the received signals. Even though non-coherent detectors
are suboptimal, they are insusceptible to dispersive effects of the channel, and therefore
well-suited for communication with implants.

The simplest non-coherent receiver concept is an energy detector, which basically consists
of a squaring device and an integrator (see Sec. 2.3). Besides its simplicity, the advantage of
an energy detector is that possible narrow-band interferers within the operating frequency
band could be suppressed by using the comb filter approach described in Sec. 3.2. This
would improve the SNR additionally. Hence, the energy detector is chosen as the receiver
in the demonstration setup for communication with implants. In this demonstration, on-off
keying is used as the modulation scheme and a unidirectional transmission with a data rate of
100 Mbit/s is selected. This data rate and a unidirectional transmission are sufficient for most
medical applications.

The transmitter in Fig. 50 consists of a UWB impulse generator (IG) and an antenna. The
transmit data are generated externally by a bit pattern generator, clocked by a signal generator.
The receiver topology is the same as shown in Sec. 2.3.2, but in a single-ended configuration.
The output signal is amplified and displayed on an oscilloscope. For further signal processing
in the digital domain, a comparator circuit with a properly adjusted threshold voltage can be
applied. A more detailed description of the transmitter and receiver structures used can be
found in [21].

To demonstrate the transmission in an environment similar to human tissue with a highly
dispersive and lossy behavior, two antennas are immersed in a tissue-mimicking liquid. This
liquid consists mainly of sugar and water. The properties of the liquid are similar to skin tissue
with a relative dielectric constant ε_r of 28 at 7 GHz and an attenuation of about 22 $\frac{dB}{cm}$. The
influence of the dispersive behavior on the impulse shape is illustrated in Fig. 51 in time and
frequency domain. There, the output signal after the transmission through the phantom liquid
is shown in comparison to the input signal. Due to increased losses for higher frequencies, the
impulse shape is significantly broadened and the amplitude is decreased by approximately
60 dB for the used path distance of 23.5 mm between the two immersed antennas.

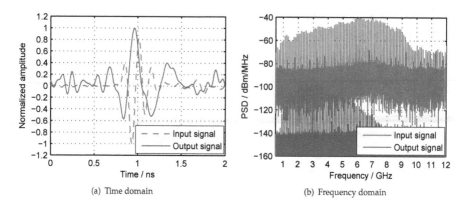

(a) Time domain   (b) Frequency domain

Figure 51. Typical received signal after the transmission of an impulse through tissue-mimicking liquid. The distance between transmitter and receiver is $R_0 = 23.5$ mm.

A data transmission is still achievable in these cases as illustrated by the reception of a typical bit pattern in Fig. 52(a). To this end, an additional amplifier with 12 dB gain is inserted after the impulse generator in order to compensate for the high losses. This measurement setup demonstrates the transmission from a deeply implanted device to a reading device placed directly on the human body, e.g. applicable for capsular endoscopy. In a final scenario, the communication of a less deeply implanted device with a base station outside the human body is considered. There, one antenna is immersed in the phantom liquid and a second one is located outside in free space (see Fig 50(b)). In this setup, the distance of the immersed antenna to free space is fixed to 15 mm, and the location of the outer antenna is varied. Similar measurement results as before are obtained here. In Fig. 52(b) the output voltage of the energy detector is plotted against the distance R_2 between the medium surface and receiver in air. The observed maximum distance of the base station for a reception is 15 cm, then the attenuation limit is reached.

4.4. Surface estimation and subsurface localization measurements

4.4.1. Measurement setup

For the evaluation of the surface estimation and subsurface localization algorithms presented in Sec. 3.4, measurements are performed using a similar bistatic UWB radar sensor as described in Sec. 4.1 [22] and the miniaturized antenna optimized for radiation in human tissue, presented in Sec. 2.1.3. The measurement setup is illustrated in Fig. 53. In this setup we use one single radar sensor which is emulating a whole sensor array by being moved along the x- and y-axis in front of the target. The target object is a container filled with the same tissue-mimicking liquid already used in the measurements of Sec. 4.3. A control unit providing clock signals for the radar sensor is connected to a computer where the signal processing and visualization is performed. In the measurements for subsurface transmitter localization, the upper elements of the setup in Fig. 53 are switched on. The miniaturized antenna placed inside of the tissue-mimicking liquid is now transmitting a 5th derivative of a Gaussian pulse generated by the impulse generator (IG). Because of the high loss in the

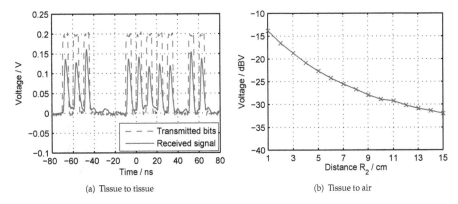

(a) Tissue to tissue   (b) Tissue to air

Figure 52. Measured output voltage of the energy detector according to the measurement scenarios in Fig. 50 using a data rate of 100 Mbit/s.

medium of about 22 $\frac{dB}{cm}$, an amplifier is used. For localization, the radar sensor outside of the container is operating in receive mode measuring the ToA of the transmitted impulse.

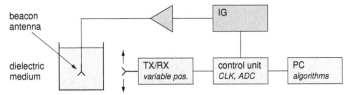

Figure 53. Measurement setup for surface estimation and subsurface transmitter localization. The radar sensor is moved in front of the liquid container to scan its surface. For localization, the upper part consisting of an impulse generator and an amplifier is switched on.

4.4.2. Surface estimation results

In 3D surface measurements the performance of the trilateration-based imaging algorithm derived in Sec. 3.4.1 is verified. As a target object a plastic dummy of a female torso of about 60 cm height as pictured in Fig. 54(a) has been chosen. In order to increase the target's reflectivity the surface of the dummy has been treated with highly conductive copper laquer. Radar scans of the 3D surface are performed by moving the radar sensor up and down in 1 cm-steps along the y-axis and by rotating the target object in 5°-steps. As a result, a cylindrical array of radar sensors is emulated.

Fig. 54(b) shows the output of the imaging algorithm, a cloud of estimated surface points representing the front side of the target. An interpolation of these points is necessary to obtain a surface model needed for the subsurface localization application. The interpolated surface is illustrated in Fig. 54(c) showing a good agreement with the original target object.

The performance of the proposed algorithm regarding its accuracy is compared to the well-established imaging algorithms "Seabed" [29] and "Envelope of Spheres" [16] by evaluating the surface measurement of a metal sphere with a known diameter of 35 cm. A

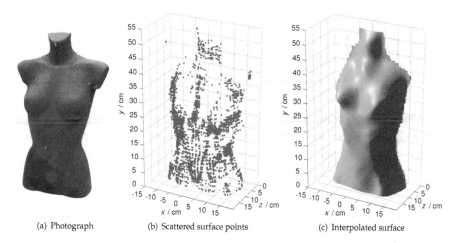

(a) Photograph   (b) Scattered surface points   (c) Interpolated surface

Figure 54. Surface estimation result of the proposed trilateration-based imaging algorithm for a radar measurement of a human torso dummy using a circular antenna arrangement.

spherical target was chosen here since a mathematical surface of the torso dummy model is unavailable. The error distance between estimated points and the ideal surface of the target sphere is calculated and plotted in Fig. 55 for different densities of measurement points. The graphs show the percentage of estimated points having a certain deviation in cm from the ideal surface. In order to compare the results of the different algorithms at exactly the same coordinates, the estimated points have been interpolated on an identical coordinate grid.

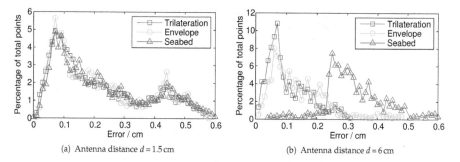

(a) Antenna distance $d = 1.5\,\mathrm{cm}$ (b) Antenna distance $d = 6\,\mathrm{cm}$

Figure 55. Percentaged distribution of the error between estimated surface points and the actual surface of a trapezoidal target using a monostatic radar setup with two different distances d between measuring points. The performance of the proposed algorithm is compared to state-of-the-art imaging algorithms.

While in the first measurement in Fig. 55(a) a relatively small step width of $d = 1.5\,\mathrm{cm}$ between two measuring positions has been used, the measurement results in Fig. 55(b) show the errors obtained with a quadrupled step width of 6 cm. It can be seen that with a high density of measurement points there is no significant difference of estimation errors between the three compared algorithms. The deviations from the ideal surface points are in a low millimeter range. However, with an increased distance between measurement points the performance of "Seabed" is degrading significantly, while the errors obtained with trilateration and the "Envelope of Spheres" algorithm remain in the same range.

It can be seen that the trilateration-based imaging algorithm achieves similar results as state-of-the-art surface estimation algorithms while the compexity is reduced since here no preprocessing of measurement data is needed. A more detailed description of the trilateration-based imaging algorithm and further measurement results can be found in [26].

4.4.3. Transmitter localization results

In the measurements for the evaluation of 3D transmitter localization, a liquid container whose concave surface roughly approximates the surface of a human body is chosen. Fig. 56 shows a top and side view of the measuring setup with the transmitter placed in a glass fish bowl of 21 cm diameter. The radar sensor is scanning in 1 cm steps in x- and y-direction emulating an antenna array with 14×11 elements in total.

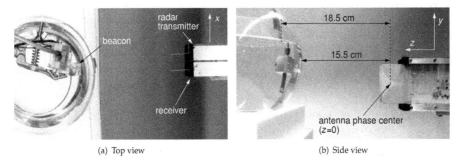

(a) Top view (b) Side view

Figure 56. Photographs of the 3D measurement setup with a transmitter placed in a concave liquid container.

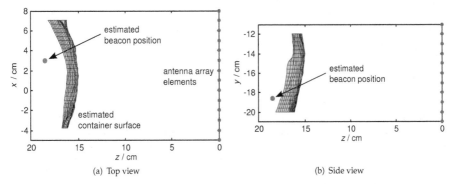

(a) Top view (b) Side view

Figure 57. Measurement results in a top and side view corresponding to the photographs in Fig. 56 showing the estimated location of the beacon behind the container surface. The dots at $z = 0$ represent the sensor positions.

The surface points found by the trilateration-based imaging algorithm are interpolated to create the boundary shape depicted in Fig. 57. These graphs follow the perspectives of the photographs in Fig. 56 allowing a rough evaluation of the localization result. By applying the localization method of Sec. 3.4.2 based on the evaluation of 3D wavefronts inside of the medium, an estimated position of the transmitter of $(x,y,z) = (3.0\,\text{cm}, -18.4\,\text{cm}, 18.5\,\text{cm})$ is calculated. Even though an exact verification of the antenna position inside of the liquid is difficult, the manually measured z-distance of 18.5 cm between the transmitter and the receiver agrees with the estimated beacon position. In the photographs of Fig. 56 the receiver antenna is positioned at the x- and y-position closest to the beacon antenna in the container. These known sensor coordinates of (3 cm, -19 cm) also coincide well with the estimated beacon position. In case of a convex surface like the fish bowl an even better localization can be expected when using a circular or spherical arrangement of the receivers around the medium, instead of a planar arrangement.

Due to the high attenuation of signals transmitted through tissue-mimicking liquid we can only localize positions close to the container surface. In our measurements the signal-to-noise ratio of impulses running through more than 3 cm of tissue-mimicking liquid became too low to be detected. One possible way to increase the maximum transmitter distance from the surface would be the use of a cascade of multiple amplifiers to realize a higher transmitter power output. Regarding the optimization of signal processing an approach based on compressed sensing is investigated to cope with lower signal-to-noise ratios at the receiver [34].

5. Conclusion

In this chapter, novel hardware components, algorithms, systems and possible approaches in the field of UWB radar and communications for biomedical applications have been presented.

Three novel UWB antenna concepts have been introduced targeting different tasks in the medical field. For communication purposes, a circular slot antenna excited with a dipole element has been presented. Its novel differential feeding concept suppresses parasitic radiation by cable currents on the feed lines. In a new concept for applications requiring directive antennas with small beamwidths, a dielectric rod has been added to the circular slot antenna, resulting in a compact and easy-to-fabricate antenna with a high mean gain of 8.7 dBi. Furthermore, a miniaturized UWB slot antenna, optimized for the radiation in human tissue has been designed.

A flexible, differential chipset using Si/SiGe HBT technology for IR-UWB applications has been presented. On the transmit side, a low power impulse generator based on cross-coupled LC oscillator is successfully realized. It generates ns-duration and stable impulses with a spectrum well fitting the FCC mask. This impulse generator has been successfully extended to include tunability to the FCC, ECC and Japanese UWB masks as well as a biphase modulation function. On the receive side, an energy detection receiver optimized for simple on-off keying communications and a correlation detection receiver for short range radar applications have been presented. Both receivers are based on a fully differential UWB low-noise amplifier and a four-quadrant RF multiplier which performs either squaring or multiplication operations.

Using the aforementioned components, two bistatic radar systems, a single-ended and a differential configuration, have been built. Their performance has been demonstrated in a setup for vital sign detection.

Regarding UWB communications, a proposal for a transmission scheme has been discussed, using a special spread spectrum method and energy detection combined with a comb filter, which improves the SNR and rejects narrowband interference. The robustness of this concept has been demonstrated for multipath propagation channels as well as for narrowband interference, noise, and synchronization errors. The approach fits well to medical applications, because small multipath delay spreading promises an easier realization of the analog time delay needed for the comb filter. A trade-off between the number of UWB impulses per symbol (bit) and the data rate requirement has to be made for different applications.

In addition, a new concept for an impulse-radio transmission based on code shift keying with a comb filter receiver has been introduced. In this concept, the delay of the analog delay element could be shorter than the channel impulse response. It has been shown and verified by simulation that the performance and the resistance against multipath propagation, noise, narrow band and multisensor/multiuser interference are the same as in the original approach with longer delays in the comb filter loop.

Simulation results have shown that particle filtering can improve the ranging and tracking performance of an impulse UWB radar substantially in scenarios with low signal-to-noise ratio and cluttering in comparison with more conventional methods. A trade-off between realization complexity and performance can be adjusted thanks to the flexibility of the proposed algorithm.

It has been shown practically that a non-coherent energy detector is a suitable receiver concept for UWB communication with implants. The energy detector operates without synchronization and is insusceptible to dispersive effects of the channel. Demonstrational measurements in tissue-mimicking liquid have been performed with a data rate of 100 Mbit/s meeting the requirements for modern medical devices.

A 3D surface estimation algorithm based on trilateration for ultra-wideband pulse radars has been presented and derived mathematically. Since this method needs no preprocessing of measurement data its implementation is very simple. In 3D surface measurements the performance of the proposed algorithm has been verified, and comparisons with established algorithms have shown a similar performance regarding estimation errors.

As a next step towards the targeted application of catheter tracking, a method for the localization of UWB transmitters buried in homogeneous dielectric media has been presented. With the aid of surface estimation algorithms a localization behind an arbitrarily shaped medium boundary is possible. For this purpose we have proposed a system consisting of an array of UWB radar sensors outside the medium and a beacon inside the medium transmitting a short UWB pulse. The external sensors serve for surface scanning and for measuring the time of arrival of the transmitted signal. The performance of the proposed localization algorithm has been verified using electromagnetic field simulations and measurements, in which a transmitter has been placed in tissue-mimicking liquid.

Author details

Dayang Lin, Michael Mirbach, Thanawat Thiasiriphet, Jürgen Lindner, Wolfgang Menzel and Hermann Schumacher
Ulm University, Germany

Mario Leib
EADS Deutschland GmbH, Division Cassidian, Ulm, Germany

Bernd Schleicher
TriQuint Semiconductor GmbH, Munich, Germany

6. References

[1] Ahmad, F. & Amin, M. G. [2006]. Noncoherent approach to through-the-wall radar localization, *IEEE Transactions on Aerospace and Electronic Systems* 42(4): 1405–1419.

[2] Belostotski, L. & Haslett, J. [2009]. A technique for differential noise figure measurement with a noise figure analyzer, *IEEE Microwave Magazine* pp. 158–161.

[3] Bilich, C. G. [2006]. UWB radars for Bio-Medical Sensing: Attenuation Model for Wave Propagation in the body at 4GHz, *University of Trento Technical report DIT-06-051* .

[4] Blech, M. D. & Eibert, T. F. [2007]. A Dipole Excited Ultrawideband Dielectric Rod Antenna With Reflector, *IEEE Transactions on Antennas and Propagation* 55(7): 1948–1954.

[5] Buchegger, T., Ossberger, G., Reisenzahn, A., Hochmaier, E., Stelzer, A. & Springer, A. [2005]. Ultra-Wideband Transceivers for Cochlear Implants, *EURASIP Journal on Applied Signal Processing* 18: 3069–3075.

[6] CST [2011]. *CST MICROWAVE STUDIO®, User Manual*. Darmstadt, Germany.

[7] Dederer, J. [2009]. *Si/SiGe HBT ICs for Impulse Ultra-Wideband (I-UWB) Communications and Sensing*, Cuvillier Verlag, Göttingen, Germany.

[8] Dederer, J., Schleicher, B., Santos, F., Trasser, A. & Schumacher, H. [2007]. *FCC compliant 3.1-10.6 GHz UWB Pulse Radar using Correlation Detection*, Proc. IEEE MTT-S Int. Microw. Symp., Honolulu, HI, pp. 1471–1474.

[9] Dederer, J., Schleicher, B., Trasser, A., Feger, T. & Schumacher, H. [2008]. A Fully Monolithic 3.1-10.6 GHz UWB Si/SiGe HBT Impulse-UWB Correlation Receiver, *IEEE International Conference on Ultra-Wideband (ICUWB)*, Vol. 1, pp. 33–36.

[10] Devine, P. [2000]. *Radar Level Measurement: The User's Guide*, VEGA Controls.

[11] Fear, E. C., Hagness, S. C., Meaney, P. M., Okoniewski, M. & Stuchly, M. A. [2002]. Enhancing Breast Tumor Detection With Near-field Imaging, *IEEE Microwave Magazine* 3(1): 48–56.

[12] Fort, A., Desset, C., Ryckaert, J., Doncker, P. D., Biesen, L. V. & Wambacq, P. [2005]. Characterization of the ultra wideband body area propagation channel, *Ultra-Wideband, 2005. ICU 2005. 2005 IEEE International Conference on*, p. 6 pp.

[13] Gabriel, S., Lau, R. W. & Gabriel, C. [1996]. The dielectric properties of biological tissues: II. Measurements in the frequency range 10 Hz to 20 GHz, *Phys. Med. Biol.* 41: 2251.

[14] Hees, A., Hasch, J. & Detlefsen, J. [2008]. Characteristics of a Corrugated Tapered Slot Antenna with Dielectric Rod and Metallic Reflector, *IEEE International Conference on Ultra-Wideband*, Vol. 1.

[15] Hertel, T. W. [2005]. Cable-current Effects of Miniature UWB Antennas, *IEEE International Symposium on Antennas and Propagation*, Vol. 3A, pp. 524–527.

[16] Kidera, S., Sakamoto, T. & Sato, T. [2009]. High-Resolution 3-D Imaging Algorithm With an Envelope of Modified Spheres for UWB Through-the-Wall Radars, *Antennas and Propagation, IEEE Transactions on* 57(11): 3520–3529.

[17] Kidera, S., Sakamoto, T. & Sato, T. [2010]. Accurate UWB Radar Three-Dimensional Imaging Algorithm for a Complex Boundary Without Range Point Connections, *Geoscience and Remote Sensing, IEEE Transactions on* 48(4): 1993–2004.

[18] Krach, B., Lentmaier, M. & Robertson, P. [2008]. Joint bayesian positioning and multipath mitigation in gnss, *Acoustics, Speech and Signal Processing, 2008. ICASSP 2008. IEEE International Conference on*, pp. 3437–3440.

[19] Leberer, R. [2005]. *Untersuchung von quasi-planaren Antennen mit sektorförmiger und omnidirektionaler Strahlungscharakteristik im Millimeterwellenbereich*, Dissertation, Universität Ulm, Institut für Mikrowellentechnik.

[20] Leib, M. [2011]. *Ultrabreitbandige Antennen für Kommunikation und Sensorik in der Medizintechnik*, Dissertation, Universität Ulm, Institut für Mikrowellentechnik.

[21] Leib, M., Mach, T., Schleicher, B., Ulusoy, C. A., Menzel, W. & Schumacher, H. [2010]. Demonstration of UWB communication for implants using an energy detector, *German Microwave Conference, 2010*, pp. 158–161.

[22] Leib, M., Schmitt, E., Gronau, A., Dederer, J., Schleicher, B., Schumacher, H. & Menzel, W. [2009]. A compact ultra-wideband radar for medical applications, *Frequenz* 63(1-2): 2–8.

[23] Lin, D., Trasser, A. & Schumacher, H. [2011a]. *A Fully Differential IR-UWB Front-end for Noncoherent Communication and Localization*, Proc. IEEE ICUWB 2011, Bologna, Italy, pp. 116–120.

[24] Lin, D., Trasser, A. & Schumacher, H. [2011b]. *Low Power, Fully Differential SiGe IR-UWB Transmitter and Correlation Receiver ICs*, Proc. IEEE RFIC, Baltimore, MD, pp. 101–104.

[25] Merdes, C. L. & Wolf, P. D. [2001]. Locating a catheter transducer in a three-dimensional ultrasound imaging field, *Biomedical Engineering, IEEE Transactions on* 48(12): 1444–1452.

[26] Mirbach, M. & Menzel, W. [2011]. A simple surface estimation algorithm for UWB pulse radars based on trilateration, *ICUWB 2011, 2011 IEEE International Conference on Ultra-Wideband*, pp. 273–277.

[27] Mirbach, M. & Menzel, W. [2012]. Time of Arrival Based Localization of UWB Transmitters Buried in Lossy Dielectric Media, *ICUWB 2012, 2012 IEEE International Conference on Ultra-Wideband*, pp. 328–332.

[28] Rappaport, C. [2004]. A simple approximation of transmitted wavefront shape from point sources above lossy half spaces, *Geoscience and Remote Sensing Symposium, 2004. IGARSS '04. Proceedings. 2004 IEEE International*, Vol. 1, pp. 421–424.

[29] Sakamoto, T. [2007]. A fast algorithm for 3-dimensional imaging with UWB pulse radar systems, *IEICE Transactions on Communications* E90-B(3): 636–644.

[30] Schantz, H. [2005]. *The Art and Science of Ultra-Wideband Antennas*, Artech House, Norwood, USA.

[31] Schleicher, B. [2012]. *Impulse-Radio Ultra-Wideband Systems for Vital-Sign Monitoring and Short-Range Communications*, Verlag Dr. Hut, München, Germany.

[32] Schueppert, B. [1988]. Microstrip/Slotline Transitions: Modeling and Experimental Investigation, *IEEE Transactions on Microwave Theory and Techniques* 36(8): 1272–1282.

[33] Staderini, E. M. [2002]. UWB Radars in Medicine, *IEEE Aerospace and Electronic Systems Magazine* 17(1): 13–18.

[34] Thiasiriphet, T., Ibrahim, M. & Lindner, J. [2012]. *Compressed Sensing for UWB Medical Radar Applications*.

[35] Thiasiriphet, T. & Lindner, J. [2010]. An uwb receiver based on a comb filter with shortened delay, *Ultra-Wideband (ICUWB), 2010 IEEE International Conference on*, Vol. 1, pp. 1–4.

[36] Thiasiriphet, T. & Lindner, J. [2011]. Particle filtering for uwb radar applications, *Ultra-Wideband (ICUWB), 2011 IEEE International Conference on*, pp. 248–252.

[37] Thiasiriphet, T., Zhan, R. & Lindner, J. [2009]. Interference suppression in ds-ppm uwb systems, *Ultra-Wideband. ICUWB 2009. 2009 IEEE International Conference on*, pp. 718–722.

[38] Thiel, F., Hein, M., Schwarz, U., Sachs, J. & Seifert, F. [2008]. Fusion of Magnetic Resonance Imaging and Ultra-Wideband-Radar for Biomedical Applications, *IEEE International Conference on Ultra-Wideband*, Vol. 1, pp. 97–100.

[39] Zheng, Y., Wong, K., Asaru, M., Shen, D., Zhao, W., The, Y., Andrew, P., Lin, F., Yeoh, W. & Singh, R. [2007]. *A 0.18um CMOS Dual-Band UWB Transceiver*, IEEE ISSCC Dig. Tech. Papers, San Francisco, CA, pp. 114–115.

[40] Zhou, Z., Pan, Z. & Tang, X. [2007]. A new family of optimal zero correlation zone sequences from perfect sequences based on interleaved technique, *Signal Design and Its Applications in Communications, 2007. IWSDA 2007. 3rd International Workshop on*, pp. 195–199.

Permissions

The contributors of this book come from diverse backgrounds, making this book a truly international effort. This book will bring forth new frontiers with its revolutionizing research information and detailed analysis of the nascent developments around the world.

We would like to thank Reiner Thomä, Reinhard Knöchel, Jürgen Sachs, Ingolf Willms and Thomas Zwick, for lending their expertise to make the book truly unique. They have played a crucial role in the development of this book. Without their invaluable contribution this book wouldn't have been possible. They have made vital efforts to compile up to date information on the varied aspects of this subject to make this book a valuable addition to the collection of many professionals and students.

This book was conceptualized with the vision of imparting up-to-date information and advanced data in this field. To ensure the same, a matchless editorial board was set up. Every individual on the board went through rigorous rounds of assessment to prove their worth. After which they invested a large part of their time researching and compiling the most relevant data for our readers. Conferences and sessions were held from time to time between the editorial board and the contributing authors to present the data in the most comprehensible form. The editorial team has worked tirelessly to provide valuable and valid information to help people across the globe.

Every chapter published in this book has been scrutinized by our experts. Their significance has been extensively debated. The topics covered herein carry significant findings which will fuel the growth of the discipline. They may even be implemented as practical applications or may be referred to as a beginning point for another development. Chapters in this book were first published by InTech; hereby published with permission under the Creative Commons Attribution License or equivalent.

The editorial board has been involved in producing this book since its inception. They have spent rigorous hours researching and exploring the diverse topics which have resulted in the successful publishing of this book. They have passed on their knowledge of decades through this book. To expedite this challenging task, the publisher supported the team at every step. A small team of assistant editors was also appointed to further simplify the editing procedure and attain best results for the readers.

Our editorial team has been hand-picked from every corner of the world. Their multi-ethnicity adds dynamic inputs to the discussions which result in innovative outcomes. These outcomes are then further discussed with the researchers and contributors who give their valuable feedback and opinion regarding the same. The feedback is then collaborated with the researches and they are edited in a comprehensive manner to aid the understanding of the subject.

Apart from the editorial board, the designing team has also invested a significant amount of their time in understanding the subject and creating the most relevant covers. They scrutinized every image to scout for the most suitable representation of the subject and create an appropriate cover for the book.

The publishing team has been involved in this book since its early stages. They were actively engaged in every process, be it collecting the data, connecting with the contributors or procuring relevant information. The team has been an ardent support to the editorial, designing and production team. Their endless efforts to recruit the best for this project, has resulted in the accomplishment of this book. They are a veteran in the field of academics and their pool of knowledge is as vast as their experience in printing. Their expertise and guidance has proved useful at every step. Their uncompromising quality standards have made this book an exceptional effort. Their encouragement from time to time has been an inspiration for everyone.

The publisher and the editorial board hope that this book will prove to be a valuable piece of knowledge for researchers, students, practitioners and scholars across the globe.

List of Contributors

Henning Mextorf, Frank Daschner, Mike Kent and Reinhard Knöchel
University of Kiel, Germany

Alexander Esswein and Robert Weigel
Institute for Electronics Engineering, University of Erlangen-Nuremberg, Germany

Christian Carlowitz and Martin Vossiek
Institute of Microwaves and Photonics, University of Erlangen-Nuremberg, Germany

Ingrid Hilger, Katja Dahlke, Gabriella Rimkus and Christiane Geyer
Jena University Hospital, Germany

Frank Seifert, Olaf Kosch and Florian Thiel
Physikalisch-Technische Bundesanstalt Berlin, Germany

Matthias Hein, Francesco Scotto di Clemente, Ulrich Schwarz, Marko Helbig and Jürgen Sachs
Ilmenau University of Technology, Germany

Stefan Heinen, Ralf Wunderlich and Markus Robens
RWTH Aachen University, Germany

Jürgen Sachs and Martin Kmec
Ilmenau University of Technology, Germany

Robert Weigel, Thomas Ußmüller, Benjamin Sewiolo and Mohamed Hamouda
Friedrich-Alexander Universität Erlangen, Germany

Rolf Kraemer, Johann-Christoph Scheytt and Yevgen Borokhovych
Brandenburgische Technische Universität Cottbus, Germany

Dayang Lin, Michael Mirbach, Thanawat Thiasiriphet, Jürgen Lindner,Wolfgang Menzel and Hermann Schumacher
Ulm University, Germany

Mario Leib
EADS Deutschland GmbH, Division Cassidian, Ulm, Germany

Bernd Schleicher
TriQuint Semiconductor GmbH, Munich, Germany

Printed in the USA
CPSIA information can be obtained
at www.ICGtesting.com
JSHW011426221024
72173JS00004B/695